Plant Nonprotein Amino and Imino Acids

Biological, Biochemical, and Toxicological Properties

Plant Nonprotein Amino and Imino Acids

Biological, Biochemical, and Toxicological Properties

GERALD A. ROSENTHAL

*Thomas Hunt Morgan School of Biological Sciences
and the Graduate Center for Toxicology
University of Kentucky
Lexington, Kentucky*

1982

ACADEMIC PRESS

A Subsidiary of Harcourt Brace Jovanovich, Publishers

New York   London

Paris   San Diego   San Francisco   São Paulo   Sydney   Tokyo   Toronto

ACADEMIC PRESS, INC.
111 Fifth Avenue, New York, New York 10003

United Kingdom Edition published by
ACADEMIC PRESS, INC. (LONDON) LTD.
24/28 Oval Road, London NW1 7DX

Library of Congress Cataloging in Publication Data

Rosenthal, Gerald A.
   Plant Nonprotein Amino and Imino Acids: Biological, Biochemical, and
Toxicological Properties.

   (American Society of Plant Physiologists monograph
series)
   Bibliography: p.
   Includes index.
   1. Amino acids. 2. Amino acids--Toxicology.
3. Imino acids. 4. Botanical chemistry. 5. Plants
--Metabolism. I. Title. II. Series.
QK898.A5R67   582'.019245   82-1651
ISBN 0-12-597780-8   AACR2

PRINTED IN THE UNITED STATES OF AMERICA

82 83 84 85   9 8 7 6 5 4 3 2 1

Contents

Preface

Several reviews on higher plant nonprotein amino and imino acids, most emphasizing the toxic nature of some members, have appeared recently, and general information on these secondary plant metabolites is available. Past works, however, have a single common deficiency—the scope of their presentation is narrowed by the limitation of space. No single text exists as a reference source for these important compounds, and the literature on this topic is scattered over an unusually diversified array of journals representing many distinctive fields. Finally, no work is available for graduate level readers that can aid and direct their introduction to this area of plant science.

When the opportunity arose to author a treatise on the nonprotein amino acids of green plants, it represented a most worthwhile challenge, particularly if directed at the desired audience. I wanted to convey my own predilection for this fascinating group of natural products and to present their marked potential as experimental tools for probing fundamentally important biological questions. I wanted to accurately represent the areas of our established knowledge, to point to certain areas where deficiencies exist, and to indicate where further work and exploration are needed.

Much of this monograph is written with the neophyte in mind. This necessitated the presentation of certain basic concepts that are well-known to the advanced worker. It is understood that such material may be omitted by these readers. In creating this unusual dichotomy in the level of presentation, I was motivated by my experiences as a graduate student and the concepts and information that I struggled with. At the same time, particularly in the formation of the latter chapters, I directed my attention to the knowledgeable reader since I wanted to create a valuable basic reference source. In either case, it was not my intention to create an exhaustive coverage of the subject matter but rather to represent effectively the state of the art, to provide a helpful means of identifying and locating the pertinent literature, and to present the basic information necessary to encourage other workers to enter into the study of the nonprotein amino acids.

Early on it was decided not to consider the constituents and contributions of protistans, since these organisms have received far more attention than higher plants. In limiting this plant presentation to monerans, I have accepted modern taxonomic delineations placing the fungi in a distinctive group. Any significant examination of their fundamental biochemical reactions reinforces the basic soundness of this delineation. Hopefully, the inconvenience or deficiency created by these decisions will be balanced by the need to maintain the text size at a level where the cost can be absorbed by the young investigator. The body of information necessary to justify an integrated presentation of the material results from the efforts of an international assemblage of scientists spanning several decades of diligent effort. It is my intent to fully document their many achievements and contributions by this work.

The number and structural types of nonprotein amino acids, and particularly nonprotein amino acid application in understanding fundamentally important biological questions, has increased dramatically over the past three decades. It is my hope that this work will contribute meaningfully to the enhanced scrutiny and value that these natural products will receive over the next several decades.

Gerald A. Rosenthal

Acknowledgments

An undertaking of this nature requires the contributing efforts of many individuals so that a body of extensive information can be presented in an accurate, comprehensive, and lucid manner. I was aided significantly in this task by the incisive evaluations and suggestions of Drs. S. F. Conti and T. Gray of the T. H. Morgan School of Biological Sciences and Drs. W. Smith and S. Smith of the Chemistry Department of the University of Kentucky. Dr. E. E. Conn of the University of California at Davis; P. J. Lea of Rothamsted Experimental Station, England; John Giovanelli, National Institutes of Health; David Seigler, University of Illinois; John Thompson, USDA, Cornell University; Leonard Beevers, University of Oklahoma; and Dr. Ann Oaks, McMaster's University all graciously gave of their time and knowledge. This activity is a time-consuming, tedious, and demanding effort and one which carries no real professional reward. I can only offer my sincere appreciation and thanks for their efforts in providing this valuable service; it meant a great deal to me.

I gratefully acknowledge the financial assistance provided me in a series of grants from the National Institutes of Health, the National Science Foundation, and the University of Kentucky Research Foundation. Their support made possible my research efforts documented in this monograph. A portion of this work was completed as a Visiting Professor of Botany at Seoul National University, Seoul, Korea under the auspices of the Agency for International Development. Additional progress was made as a Lady Davis Visiting Professor of Entomology at Hebrew University of Jerusalem, Rehovot Campus, Israel. I am grateful for the financial support extended to me during this period when much of the work was achieved.

Many demanding tasks were performed by Amy-Jo Rosenthal, Carol Chambers, Darla Morrow, and Mark Schmidt; Bobbie Welch and Judi Cromer did much of the typing. A special thanks is due to my wife, Carol, for her countless hours of proofreading, general efforts, and many helpful suggestions. She continues to be a constant source of encouragement and aid in my professional activities of significance.

Chapter 1

Nomenclature and Certain Physicochemical Properties

These materials (amino acids) are at the same time substituted bases and substituted acids. Their capacity to function as amphoteric electrolytes has therefore conferred on them many remarkable electro-chemical properties not shared by any other product of natural origin.

(Greenstein and Winitz, 1961).

A. INTRODUCTION

Nitrogen is the most abundant component of Earth's atmosphere but only a handful of higher plants can contribute to the utilization of this indispensable but relatively inert element. These higher plants, by virtue of symbiotic microbial associations, possess the unique ability to fix vast quantities of diatomic nitrogen into ammonia which is toxic and has limited biological utility until it is assimilated into organic linkage. The biosynthesis of amino acids represents the principal means for the assimilation of fixed nitrogen into biologically functional molecules.

What exactly then is an amino acid? It is a substance having both an amino group and an organic acid. The labile proton can be derived not only from the customary carboxyl group but also by ionization of sulfonic acid. For practical purposes, however, the term amino acid is applied to compounds sharing the general structure $R—CH(NH_2)COOH$. While an amino group is usually linked directly to a carbon atom alpha to that of the carboxyl group, the amino group may

be associated with any carbon, such as in β-alanine, $H_2N—CH_2—CH_2—$
COOH, or γ-aminobutyric acid, $H_2N—CH_2—CH_2—CH_2—COOH$.

As a group, the nonprotein amino acids are extremely diversified and
it is not surprising that several systems can be employed for classifying
and ordering these natural products. They may be divided arbitrarily
into groups according to their structure (e.g., aliphatic, aromatic, or
heterocyclic); the number and nature of their ionizable groups; their
basic, acidic, or neutral character; their polar or apolar nature; and finally
their physiological properties and biological effects in selected organisms. In this volume, I have selected aspects of the first two group-
ings for the ordering of the nonprotein amino acid and imino acids of
plants (see the Appendix).

A very large number of plant nonprotein amino acids are saturated
aliphatic amino acids with an additional amino or carboxyl group;
only 2,6-diaminopimelic acid carries both additional groups while
4-carboxy-4-hydroxy-2-aminoadipic acid has three carboxyl groups. A
limited number of nonprotein amino acids are unsaturated aliphatics
characterized by ethylenic or acetylenic linkages, and recently pyrrolidine or cyclopropane-ring structures having an exocyclic methylene
function have been described.

Nearly all nonprotein amino acids occur in the free form but an occasional one is isolated attached to a carbohydrate moiety and a few dozen
are found as γ-glutamyl-linked peptides. Many of these compounds
exist in homologous series and bear some structural analogy to their protein amino acid counterpart.

In addition to the 20 or so universally distributed protein amino acids,
at this time over 400 others have been obtained from natural sources.[*]
About 240 nonprotein amino acids are found in various plants. Prokaryotic organisms are the source for an additional 50, while the fungi[†]
provide 75 others. Animals are not known to uniquely produce more
than about 50 kinds.

Many of these natural products are aromatic or heterocyclic in their
structure. One quarter of all nonprotein amino acids are hydroxylated
and this applies to many aromatic members; these aromatic constituents
mostly contain a phenyl group associated with alanine or glycine. The
heterocyclic nonprotein amino acids are truly varied, containing in addition to carbon either oxygen, nitrogen, or sulfur within the ring. They are
often β-substituted alanines in which pyrimidine, pyrone, pyrazole, pyri-

---

[*]The occurrence of some has not been established by isolation and rigorous chemical
characterization.

[†]In this work, the fungi are taken to represent an assemblage distinctive from protistans
and higher plants.

dine, thiazole, or isoxazoline structures are evident; a large number are amino acids constructed from azetidine, pyrrolidine, or piperidine units.

B. TRIVIAL NOMENCLATURE

The nonprotein amino acids are not only highly diverse and often quite complex but also their trivial names generally fail to bear any meaningful relationship to structural configuration. While the trivial name provided a valuable means of reference prior to structural elucidation, these designations were usually applied before the establishment of conventional rules of nomenclature and, in time, became so familiar as to ensure their continued use. The trivial name is derived usually from the generic portion of the accredited Latin binomial of the original source. For example, citrulline was coined for its isolation from the juice of *Citrullus vulgaris* (watermelon) (Wada, 1930); likewise canavanine reflects its biosynthesis and isolation from *Canavalia ensiformis* (jack bean). That the generic name need not serve as the basis for the nomenclatural designation is illustrated by the concurrent isolation of a heterocyclic nonprotein amino acid from *Lathyrus tingitanus*. One discoverer named it lathyrine while the other preferred the specific name and the term tingitanine was born. The common name of the source can also inspire new terminology; djenkolic acid is termed for the djenkol bean native to Java. Nonprotein amino acid nomenclature often bears a definitive relationship to the nature and structure of the constituent groups but trivial names still predominate in the naming of these natural compounds.

C. STEREOSPECIFICITY AND FORMAL NOMENCLATURE

A substance that rotates the plane of plane-polarized light is taken to be optically active. All amino acids having the structural designation RCH(NH$_2$)—COOH, where R is other than H, have optical rotatory power, and exist in at least two distinctive isomeric forms. One form, designated *d,* indicated that the *direction* of the optical rotation was *dextrorotatory* or clockwise. The other form, termed *l,* represented levorotatory or counterclockwise rotation of the plane of the polarized light. As such, the terms *d* and *l* are strictly *rotational notations* which, while they are of historical interest, are not part of current nomenclatural practice.

The eminent chemist Emil Fischer selected the naturally occurring form of glucose, which is dextrorotatory, as the standard for the D-configurational family; but presumably due to its lesser complexity, dextrorotatory glyceraldehyde ultimately emerged as the accepted standard for the carbohydrate series. The obvious structural similarity of serine to glyceraldehyde undoubtedly instigated its selection as the standard of reference for amino acids.

$$
\begin{array}{cc}
\text{CHO} & \text{CO}_2\text{H} \\
| & | \\
\text{H—C—OH} & \text{H—C—NH}_2 \\
| & | \\
\text{CH}_2\text{OH} & \text{CH}_2\text{OH} \\
\\
\text{D(+)- Glycer-} & \text{D-Serine} \\
\text{aldehyde} &
\end{array}
$$

Until it became technically feasible to determine the *absolute configuration* of these standard compounds, it was the accepted practice to relate optically active compounds to the above reference sources. By 1949, it had become possible to determine the absolute configuration of reference organic compounds, and it was shown that by pure chance the *standard configuration* of (+)-glyceraldehyde corresponded to its experimentally determined *absolute configuration*. The absolute configuration of the α-carbon of an amino acid is taken directly from its formal relationship to L- or D-serine which are the accepted standards for deciding relevant questions of amino acid configuration. Ultimately, however, it is the absolute configuration of D- or L-glyceraldehyde which is the determinant factor, since the absolute configuration of the *enantiomeric* forms of serine are derived from this carbohydrate standard.

As mentioned previously, all amino acids, except glycine, exist in two isomeric forms created by the asymmetrical nature of the α-carbon atom. The prefixes L or D are configurational notations which reveal the chirality of a given enantiomorphic form. In stating that *enantiomorphs* are *chiral* molecules, it is meant that they possess "handedness," i.e., they are related to each other in the sense that the right hand is to the left. Thus, while enantiomorphs are related it is only so in the sense of an object's relationship to its non-superimposable mirror image. As such, they indicate the absolute configuration of the amino acid's α-carbon atom and differentiate between the two possible isomeric forms created by the chiral α-carbon atom. This relationship is enantiomorphic and the compounds are known as enantiomorphs, enantiomers, antimers, or optical antipodes (Fig. 1).

The prefixes D or L are placed before the parent compound and preceded by a hyphen; thus, L-indospicine or D-ornithine. An optically inactive mixture created by an equal number of chiral molecules and their optical antipodes, i.e., a racemic mixture, is designated by the prefix DL (*no* commas) or the notation (±). A molecule whose optical

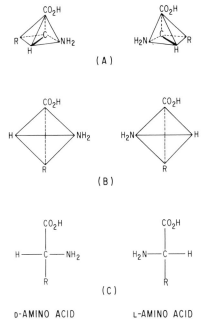

(A)

(B)

D-AMINO ACID   L-AMINO ACID

Fig. 1. Dimensional representations of an amino acid and its optical antipode. The typical tetrahedral representation is shown in (A). These enantiomers can also be represented as in (B) or in the more conventional Fischer projection (C).

rotatory power is compensated internally, thereby rendered optically inactive, is known as the *meso* form and is designated as *meso* or *ms-*, e.g., *meso*-cystine.

In naming the secondary metabolites, it is required that the substituted radicals be numbered according to their position on the carbon skeleton. The accepted designation for an aliphatic amino acid is

$$\underset{\epsilon\quad\delta\quad\gamma\quad\beta\quad\alpha}{\overset{6\quad5\quad4\quad3\quad2\qquad1}{H_2NCH_2CH_2CH_2CH_2CH(NH_2)COOH}}$$

The nomenclature of aromatic compounds, as suggested by IUPAC convention, is also facilitated by examining the numbering system of their parent compounds.

The position of the substituent is denoted by utilizing the numerical locants of the carbon skeleton, but many nonprotein amino acids are known by their Greek locants. In the latter system, the carboxyl group attached to the α-carbon is not designated specifically, because it is not possible to substitute on this particular carbon atom since it does not possess a replaceable hydrogen atom. From these considerations, it is evident that substitution of the hydrogen on carbon 4 of proline by a

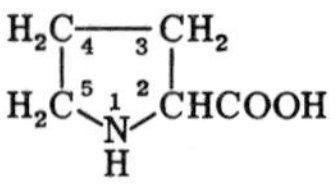

$$H_2C \underset{5}{\overset{4}{\rule{1.2em}{0.4pt}}} {\underset{}{\overset{3}{}}} CH_2$$

Proline

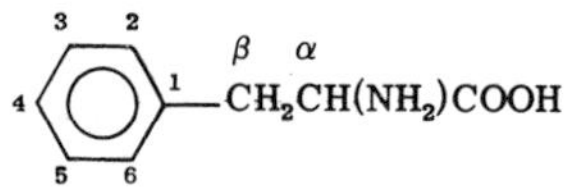

Phenylalanine

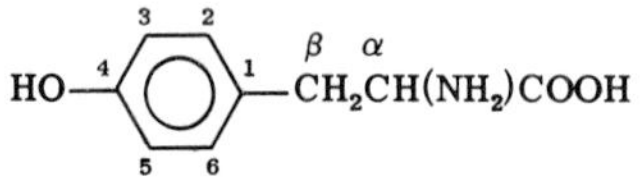

$$\beta \quad \alpha$$
$$-CH_2CH(NH_2)COOH$$

Tyrosine

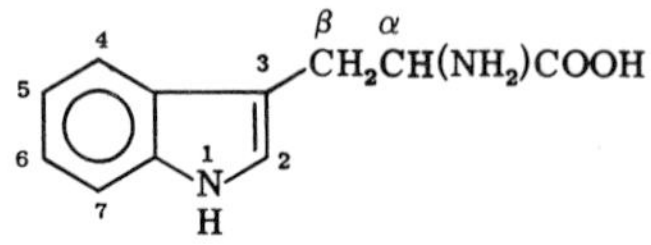

$$\beta \quad \alpha$$
$$HO- \quad -CH_2CH(NH_2)COOH$$

Tryptophan

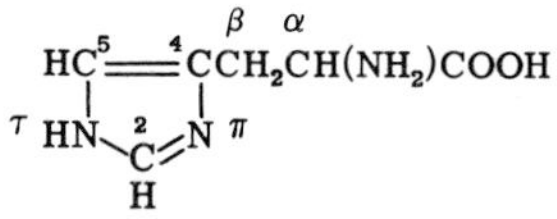

$$\beta \quad \alpha$$
$$-CH_2CH(NH_2)COOH$$

Histidine

$$\overset{\beta \quad \alpha}{HC^5 = {}^4CCH_2CH(NH_2)COOH}$$
$$^7HN \underset{C}{\overset{2}{\diagdown}} N \, \pi$$
$$H$$

hydroxyl radical produces 4-hydroxyproline. An aliphatic nonprotein amino acid such as β-aminoalanine can be referred to in a more systematic sense as α,β-diaminopropionic acid after the Greek system or 2,3-diaminopropionic acid after the numerical notation. In this particular case, the Greek notation has emerged as the commonly utilized basis for its nomenclature.

When the absolute configuration of the α-carbon is known, and this is the only chiral center, the specific configuration of the nonprotein amino acid can be described completely with the D or L configurational notation. However, nonprotein amino acids can have more than one center of chirality. More simply stated, they possess additional asymmetric carbon atoms able to generate additional sources of optical activity. (All asymmetric carbon atoms do not necessarily generate optically active compounds.) The notation 4-hydroxy-L-proline does not provide precise information on the nature of the configuration of the hydroxyl chiral center. It is evident from this discussion that while the position of atoms in the α-chiral center may be known, the exact nature of an additional center cannot be inferred simply from the α-configuration. The nature of each chiral center must be determined by appropriate methodology. Moreover, the convention of placing the L or D prefix immediately before the name of the parent amino acid reinforces the fact that the absolute configuration of the α-chiral center is known, even though other centers may not, as yet, have been determined. Thus, 4-hydroxy-L-glutamic acid rather than L-4-hydroxyglutamic acid is correct. Exceptions to this rule are permitted in the case of particularly well-known compounds such as L-hydroxyproline or L-hydroxylysine but only where the position of the substitution is known widely due to past usage.

In the case of 4-hydroxy-L-proline, in relation to a Fischer projection formula, the OH group relative to the position of the carboxyl group may reside in either the *trans* or *cis* position. When these groups are in the *trans* position, it can be designated as the *erythro* form of the compound but it can be named *trans*-4-hydroxy-L-proline or *erythro*-4-hydroxy-L$_S$-proline. When these groups reside in the *cis* position, the compound can be taken to be the *threo* form (also designated allo). These considerations make possible *cis*-4-hydroxy-L-proline, allo-4-hydroxy-L-proline, or even *threo*-4-hydroxy-L$_S$-proline.*

Since these terms are taken directly from carbohydrate nomenclature, by analogy with threose and erythrose, one must be more exacting in specifying the nature of the absolute configuration of the compound. This is achieved with the subscripts s and G which refer back to the absolute configuration of serine and glyceraldehyde, respectively.

The need for the subscripts s and G in naming certain nonprotein amino acids is better appreciated when considering the question of the configuration of (+)-tartaric acid, which possesses more than one chiral center.

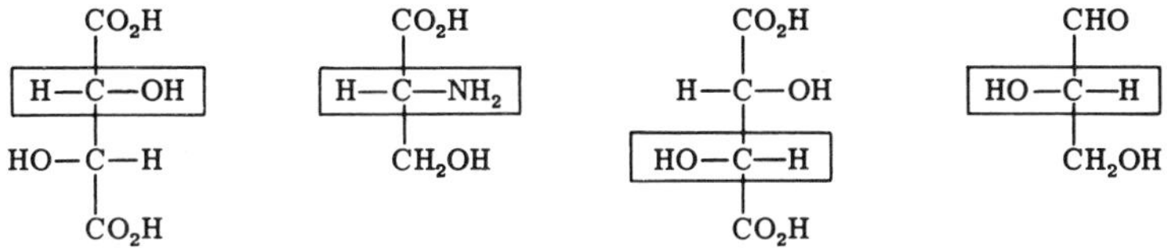

(+)-Tartaric acid   D-Serine   (+)-Tartaric acid   L-Glyceraldehyde

A   *B*

If this compound is named according to accepted rules of amino acid nomenclature, the *lowest* numbered asymmetric carbon, the carbon alpha to the carboxyl group, is employed to establish configuration. Of course, this assumes that more than one chiral center exists in the molecule. In this instance, (+)-tartaric acid's absolute configuration would be based on serine and designated D$_S$-(+)-tartaric acid. The obvious steric relationship of (+)-tartaric acid to D-serine is shown above in *A*.

*The designations of *erythro* and *threo* are the preferred nomenclature forms.

In contrast, accepted convention for naming carbohydrates dictates that the configuration of the *highest* numbered asymmetric carbon is the reference standard for determining the overall configuration of the compound. Inspection of **B** above reveals that from carbohydrate nomenclature, predicated on glyceraldehyde, (+)-tartaric acid's absolute configuration is termed L_G-(+)-tartaric acid. This situation applies to amino acids as well; for example, L-threonine can be named by either amino acid- or carbohydrate-based nomenclature rules. The subscripts s and G clearly denote which of these systems has been employed in naming the compound and avoids ambiguity in specifying absolute configuration.

When an amino acid is discovered with only a single chiral center at the α-carbon, it is relatively easy to determine its absolute configuration. For example, one need only determine if it is a substrate for appropriate stereospecific enzymes. The discovery of a second asymmetric center, however, immediately generates a compound existing in four distinctive stereoisomeric forms. Two of these species form a racemate of optical antipodes, bearing an image–non-superimposable mirror image relationship. The remaining isomers, designated the *allo* forms, are not *enantiomeric* to the first pair, i.e., they are not related as an object and its nonsuperimposable mirror image but rather share a *diastereomeric* relationship. In other words, a given isomer of an amino acid may have numerous diastereomers, but only one enantiomorphic relationship exists. In contrast to optical antipodes, diastereomers may have distinctive physical properties including magnitude of the optical rotation, solubility, infrared spectrum, and melting point. The chemical reaction rates of these isomeric forms may also be quite different.

Until the absolute configuration of the second chiral center is determined, the diastereoisomer is given the same trivial name as the parent compound but with the prefix "allo." These diastereoisomers possess the same configuration as the parent amino acid at the α-carbon atom but the opposite configuration at the second chiral center. The use of the term "allo" is discontinued once the absolute configuration of the second chiral center is established.

Recent discovery of ethylenic and acetylenic nonprotein amino acids necessitates some consideration of the procedures for naming these natural products. *Euphorbia longan* actually produces a seven-carbon saturated aliphatic acid possessing an α-amino group which is thus properly called 2-aminoheptanoic acid after the IUPAC system.

$$CH_3\!-\!CH_2\!-\!CH_2\!-\!CH_2\!-\!CH_2\!-\!CH(NH_2)COOH$$

The structure of an ethylenic compound is indicated by changing the "a" to "e" and inserting the appropriate numerical designation to indicate the position of the double bond. Thus, 2-aminohept-4-enoic acid aptly describes

$$H_3C—CH_2—CH=CH—CH_2—CH(NH_2)COOH$$

Other substituted groups are handled simply by denoting their locants on the carbon chain; for example, 2-amino-4-hydroxyhept-6-enoic acid of *E. longan*:

$$H_2C=CH—CH_2—CH(OH)—CH_2—CH(NH_2)COOH$$

The longest possible aliphatic chain should be denoted, so that the following compound

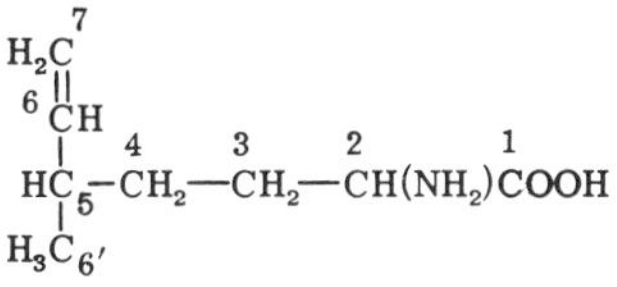

$$\overset{7}{H_2}C$$
$$\underset{6}{‖}$$
$$\overset{6}{C}H$$
$$\underset{5}{H}\overset{4}{C}—\overset{3}{C}H_2—\overset{2}{C}H_2—\overset{1}{C}H(NH_2)COOH$$
$$H_3C_{6'}$$

is correctly named 2-amino-5-methylhept-6-enoic acid and not 2-amino-5-ethylidenehexanoic acid or 2-amino-5-vinylhexanoic acid. When these factors are considered together, a designation such as 2-amino-4-hydroxyhept-6-enoic acid precisely locates and identifies the substituted groups, the position of the double bond, and the total chain length. In naming amino acids, the numbering always commences with the carboxyl group associated with the α-carbon atom since this carboxyl group is highest in priority in the naming of polyfunctional compounds. Additionally, the longest chain including the carboxyl group should be specified. The same principles apply to naming acetylenic members except that the characteristic bond is denoted by changing the "a" to "y." The acetylenic compound of *E. longan* with the structure

$$HC≡C—CH_2—CH(OH)—CH_2—CH(NH_2)COOH$$

is designated 2-amino-4-hydroxyhept-6-ynoic acid.

As an inspection of the Appendix reveals, amino acid nomenclature is a hodgepodge of mixed trivial, semisystematic, and systematic designations. A single compound can be and is known by a variety of names; consider, for example, the following substituted saturated aliphatic amino acid:

$$H_3C—CH(OH)—CH_2—CH(NH_2)COOH$$

It could conceivably be described as γ-hydroxynorvaline, 4-hydroxy-norvaline, 2-amino-4-hydroxyvaleric acid, α-amino-γ-hydroxypentanoic acid, α-amino-γ-hydroxyvaleric acid, 2-amino-4-hydroxypentanoic acid, or γ-hydroxy-α-aminovaleric acid. Greek and numerical locants are interchanged while common or IUPAC system rules are used to designate the aliphatic chain. In regard to the aliphatic chain, it is common practice to name a saturated aliphatic amino acid whose chain consists of three carbon atoms by the term propionic acid; thus, 3-hydroxy-2-aminopropionic acid or 2-amino-3-hydroxypropionic acid. A similar compound possessing a four-carbon atom-containing chain would commonly be termed 3-hydroxy-2-aminobutyric acid. The addition of a fifth carbon atom adds greater variety since these compounds can and are named from valeric acid, pentanoic acid, or by falling back to the more familiar protein amino acid root, i.e., valine, and naming it nor-valine (this reflects the linear nature of the aliphatic chain). In like manner, a sixth carbon atom generates compounds named from caproic acid, hexanoic acid, or via the protein amino acid—norleucine. In the latter instances, it is most common to employ the protein amino acid in compound nomenclature, which would probably be termed 3-hydroxynorleucine. A simple substitution on the nitrogen atom escalates a series of notations including: β-N-acetyl, N^β-acetyl, N^3-acetyl, or far worse β-acetyl to describe acetylation of a particular nitrogen atom. It is evident that far greater standardization in naming compounds is needed (see Appendix).

Finally, the prefix "homo" is conventionally employed to name a nonprotein amino acid having one more methylene group than another α-amino acid of established designation. Thus, homocitrulline and homoserine are higher homologues of citrulline and serine, respectively.

D.   INGOLD–PRELOG–CAHN CONVENTION

In 1956, Ingold, Prelog, and Cahn (see Cahn *et al.*, 1956) developed a notational system that permits expression of a compound's absolute configuration simply and unambiguously. Simply stated, in their system, the various groups attached to the chiral carbon atom are assigned a priority number on the basis of the mass of the atom that is linked directly to the asymmetric carbon atom. Priority number assignment is quite straightforward: a higher atomic mass takes precedence over lower mass. Thus, $S > O > N > C > H$ (the notation $>$ designates precedence). When multiple atoms constitute more than one attached group, each group is examined atom by atom until a dissimilarity in mass is observed. In comparing the group

$$C \begin{array}{c} \diagup H \\ — H \\ \diagdown H \end{array} \text{ with } C \begin{array}{c} \diagup H \\ — OH, \\ \diagdown H \end{array}$$

the oxygen atom is decisive for it is the first observable dissimilarity. The CH_2OH group has precedence over CH_3 since oxygen has a greater atomic mass than hydrogen. For amino acids, the group priorities are $SH > OH > NH_2 > COOH > CH_2OH > CH_3 > H$.

It is relatively easy to apply these simple principles to establish the absolute configuration of a particular amino acid. First, the assignment of the appropriate priority number is made in accordance with the sequence rule requirement that higher atomic mass takes precedence over lower. Additionally, atoms or groups attached to the chiral carbon are numbered in ascending order of priority or precedence. This is illustrated for L-alanine.

$$
\begin{array}{c}
\text{(3)} \\
\text{COOH} \\
| \\
\text{(4) } H_2N-\overset{*}{C}-H \text{ (1)} \\
| \\
\text{CH}_3 \\
\text{(2)}
\end{array}
$$

L-Alanine

Since the hydrogen atom possesses the least mass of those atoms or groups attached to the chiral carbon (1), it is given the lowest priority number. Conversely, nitrogen with a mass greater than carbon is therefore given the highest priority number—4. The remaining two atoms attached to the asymmetric carbon are themselves carbon. Thus, it is necessary to inspect the atoms associated with each of these carbons. The methyl group of alanine involves three other atoms beside carbon but all are hydrogen. The carboxyl group, however, has an oxygen atom which has a greater mass than hydrogen; this results in assignment of a priority number of three. The priority numbers for this amino acid increase from H to CH_3, to COOH, and finally NH_2

Once priority numbers have been assigned, the molecule need only be positioned so that the atom or group with the lowest priority number projects directly behind the chiral carbon atom. The remaining groups are then inspected for their relative position. (It is very helpful at this point to construct and use a simple tetrahedral model with colored sticks.) Returning to L-alanine, this molecule can be presented as a tetrahedral arrangement in which the hydrogen atom rests behind the α-carbon and is furthest from the viewer.

Priority numbers are then assigned to the various groups associated with the α-carbon atom. A continuous course traversing from 4 to 3 and then to 2 is generated. In the case of L-alanine, it moves in a counterclockwise manner.

A counterclockwise direction of movement is denoted in sequence-rule language as *sinister* (Latin, sinister = left) and given the symbol S. A clockwise rotation of the continuous course around the chiral carbon atom is denoted in sequence-rule language as *rectus* (Latin, rectus = right) and is indicated by the symbol R. The symbols R and S, written in parentheses, are followed by a hyphen prior to the actual name of the substance; thus, (R)-glyceraldehyde or (S)-alanine.

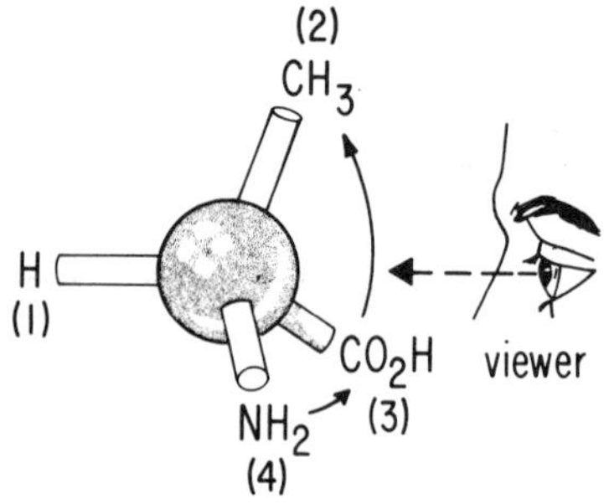

All α-amino acids have three common groups attached to the asymmetric α-carbon: H_2N, COOH, and H. If the variable R group does not take precedence over COOH, which is the case for nearly all naturally occurring protein amino acids, the orientation for a given L-isomer will always be sinister (S). A notable exception to this generalization is L-cysteine which has a sulfur atom as a substituent on the β-carbon atom. In this case, the sulfhydryl group, which is part of the β-carbon group, dictates a higher priority number than does the oxygen of the carboxyl group. Inspection of the resulting orientation of the constituent atoms and groups clearly indicates a clockwise rotation from 4 to 3 to 2. Thus, L-cysteine is also (R)-cysteine.

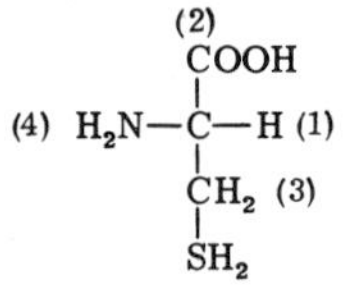

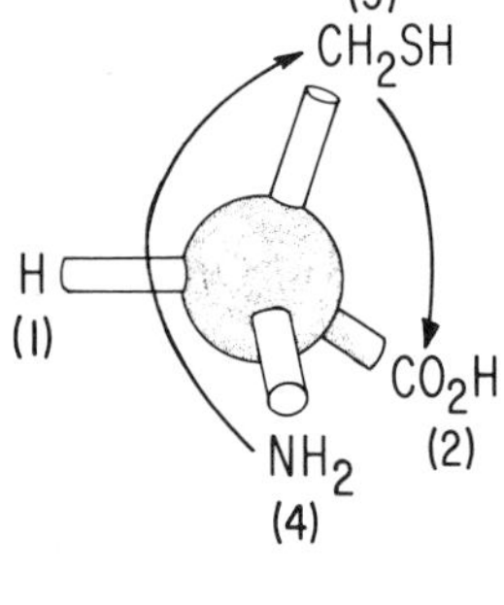

The Ingold–Prelog–Cahn convention is readily able to designate configuration when chiral centers in addition to the α-carbon exist in the molecule. In such cases, one merely applies the enumerated principles sequentially to each chiral center. Taking L_S-threonine as an example, the first chiral center (*) produces

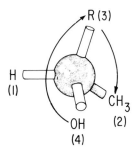

L_S-Threonine

which is counterclockwise in its rotation i.e., sinister. The second chiral center (†) generates

which is rectus. Numbering the carbon atoms from the α-carboxyl group produces $(2S,3R)$-threonine, which expresses unequivocally the absolute configuration about the asymmetric carbon atoms of the molecule. It is customary for the sequence rule symbols to be utilized for designating configuration existing at chiral centers other than the α-carbon group. The configurational terms L and D are the preferred means for indicating the nature of the α-carbon chiral center. Thus, the mixed designations are: $(3R)$-L_S-threonine and $(4S)$-4-hydroxy-L_S-proline. The sequence rule system can be used exclusively for designating both chiral centers if one desires to employ only one system. Although this admittedly simplified presentation of the Ingold–Prelog–Cahn system has been limited to a consideration of amino acids, this system simply and effectively specifies the absolute configuration for virtually all organic compounds.

E. DIPOLAR PROPERTIES

Amino acids are characteristically white, crystalline solids lacking a sharp melting or decomposition point (this severely minimizes the efficacy of this parameter in their identification); their melting or decomposition range is high, generally in excess of 200°C. Their high melting point, preferential solubility in polar rather than nonpolar solvents, and ability to elevate the dielectric constant of the solvent medium are characteristic of a molecule whose crystalline lattice consists of stabilizing electrostatic attractions between oppositely charged groups. If amino acid crystals consisted of non-ionizable groups, weaker van der Waals forces would constitute the primary stabilizing force and these natural products would have much lower melting points. This is of paramount importance since many physical properties of amino acids reflect their resemblance to electrovalent more than covalent types of compounds. Thus, rather than existing predominantly as

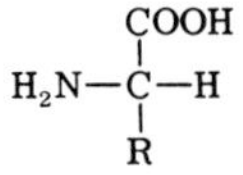

$$
\begin{array}{c}
\text{COOH} \\
| \\
\text{H}_2\text{N}-\text{C}-\text{H} \\
| \\
\text{R}
\end{array}
$$

Uncharged structure

amino acids possess an ionic structure—both in the solid state and in solution. It is customary to refer to their charged structure as a *zwitterion*. In other words, amino acids crystallize from neutral aqueous solution as charged zwitterions rather than undissociated molecules.

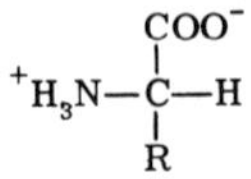

$$
\begin{array}{c}
\text{COO}^- \\
| \\
{}^+\text{H}_3\text{N}-\text{C}-\text{H} \\
| \\
\text{R}
\end{array}
$$

Zwitterion

Amino acids can function either as proton donors or acceptors.

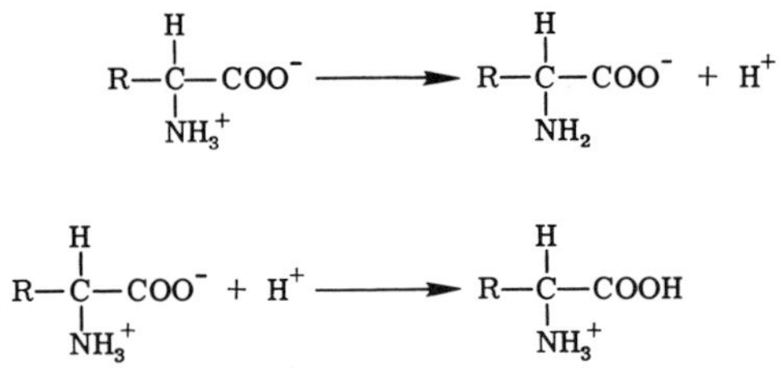

$$
\begin{array}{c}
\text{H} \\
| \\
\text{R}-\text{C}-\text{COO}^- \\
| \\
\text{NH}_3^+
\end{array}
\longrightarrow
\begin{array}{c}
\text{H} \\
| \\
\text{R}-\text{C}-\text{COO}^- + \text{H}^+ \\
| \\
\text{NH}_2
\end{array}
$$

$$
\begin{array}{c}
\text{H} \\
| \\
\text{R}-\text{C}-\text{COO}^- + \text{H}^+ \\
| \\
\text{NH}_3^+
\end{array}
\longrightarrow
\begin{array}{c}
\text{H} \\
| \\
\text{R}-\text{C}-\text{COOH} \\
| \\
\text{NH}_3^+
\end{array}
$$

The pH at which the various ionizable groups of an amino acid deprotonate can be determined experimentally. The notation pK_a, a logarithmic transformation of the apparent dissociation constant, designates the relative proton donating capacity. The pK_a values are designated in order of decreasing acidic strength. Thus, the α-COOH group, the strongest proton donor, is termed pK_{a_1}. When present, the β-COOH group usually carries the pK_{a_2} notation; the α-NH$_3^+$ group is usually termed pK_{a_3}. For aspartic acid, the respective pK_a groups are

pK_{a_1}	α-COOH	2.10
pK_{a_2}	β-COOH	3.86
pK_{a_3}	α-NH$_3^+$	9.82

These pK_a values are very useful and, as part of the Henderson–Hasselbalch equation which will be considered in greater detail later, they permit quantification of each charged group's ionization. The pK_a value represents the pH at which half of the appropriate group has deprotonated. For each amino acid there exists a pH value, known as the *isoelectric point,* at which it is electrically neutral. At this pH, the negative charge created by carboxyl or other group ionization is balanced by the positive charge generated by groups remaining fully protonated. As the pH increases, it is the α-COOH group that ionizes most readily (in aspartic acid, one-half at pH 2.1), but the other ionizable groups are simultaneously donating protons albeit at a much reduced rate. Indeed, the pH must exceed physiological levels before significant ionization of the α-NH$_3^+$ group occurs.

The concept of the isoelectric point can be developed further by examining glycine whose pK_{a_1} value = 2.34 and pK_{a_2} = 9.6. At pH 5.97, (2.34 + 9.6)/2 expresses the midpoint between the two pK_a values of glycine, the α-COO$^-$ being deprotonated to the same extent as the α-NH$_3^+$ group remains charged. The opposing electrical state but equivalent contribution of these determining groups produces the electrically neutral or isoelectric form of the amino acid. The lack of *net* charge results from the predominance of isoelectric glycine molecules and from a balance between glycine molecules that have not yet ionized the α-COOH group, and are therefore positively charged, and a small but equal number which have already deprotonated both groups. Some members of the uncharged species

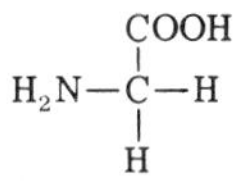

$$
\begin{array}{c}
\text{COOH} \\
| \\
\text{H}_2\text{N}-\text{C}-\text{H} \\
| \\
\text{H}
\end{array}
$$

actually exist but in infinitesimally minute amounts. At pH 5.5, for example, about one ten-millionth of the total molecular population is present as the uncharged form.

F. HENDERSON–HASSELBALCH EQUATION AND APPLICATION

A more complete consideration of the Henderson–Hasselbalch equation is desirable at this point since it provides an effective means for understanding the acid–base relationships of amino acids. This equation relates the hydrogen ion concentration to the apparent dissociation constant:

$$pH = pK_a + \log \frac{[A^0]}{[A^+]}$$

For a simple aliphatic monoamino, monocarboxylic acid, three ionization states result from the progressive deprotonation of the amino acid:

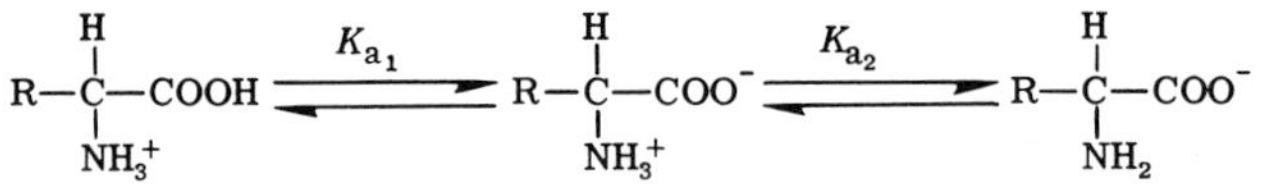

$$
\underset{A^+}{\underset{NH_3^+}{\overset{H}{R-\overset{|}{\underset{|}{C}}-COOH}}}
\xrightleftharpoons{K_{a_1}}
\underset{A^0}{\underset{NH_3^+}{\overset{H}{R-\overset{|}{\underset{|}{C}}-COO^-}}}
\xrightleftharpoons{K_{a_2}}
\underset{A^-}{\underset{NH_2}{\overset{H}{R-\overset{|}{\underset{|}{C}}-COO^-}}}
$$

K_{a_1} is the equilibrium constant for the dissociation $A^+ \rightleftharpoons A^0 + H^+$ and K_{a_2} for the reaction $A^0 \rightleftharpoons A^- + H^+$. Returning to K_{a_1} and remembering that pH is by definition the $-\log H^+$, pK_{a_1} is a similar logarithmic transformation of the apparent dissociation constant. These relationships simplify to

$$K_{a_1} = \frac{[A^0]\,[H^+]}{[A^+]}$$

$$[H^+] = K_{a_1} \frac{[A^+]}{[A^0]}$$

$$-\log[H^+] = -\log K_{a_1} -\log\frac{[A^+]}{[A^0]}$$

$$pH = pK_{a_1} + \log \frac{[A^+]}{[A^+]}$$

This equation is applicable to the dissociation of each ionizable group of the nonprotein amino acid and is valuable in understanding how various acid–base equilibria factors affect these natural products. Consider, for example, the nonprotein amino acid ornithine with the following pK_a values:

pK_{a_1}	α-COOH	1.94
pK_{a_2}	α-NH$_3^+$	8.65
pK_{a_3}	δ-NH$_3^+$	10.76

Intuitively, the isoelectric point (pI) for this compound must be at a pH where the negative charges contributed by ionization of the α-COOH group equals the sum of the positive charges contributed by both the α-and δ-NH$_3^+$ groups, that is, at a pH halfway between that of the two contributory amino groups. At this pH, the extent to which the α-NH$_3^+$ group is *deprotonated* is balanced by the δ-NH$_3^+$ group's positive charge, i.e., the extent to which it remains un-ionized. Together they counterbalance the negative charge generated by the α-COO$^-$ group. Thus, L-ornithine is isoelectric at pI $= (8.65 + 10.76)/2 = 9.70$. In another case, suppose it is desirable to obtain the following charged species of ornithine:

$$
\begin{array}{c}
\overset{+}{NH_3} \\
| \\
(CH_2)_3 \\
| \\
H-C-NH_3^+ \\
| \\
COO^-
\end{array}
$$

At what pH is this species maximized? At a pH value equal to pK_{a_1}, i.e., 1.94, although both amino groups essentially are fully charged, only one-half of the α-COOH group is ionized. As the pH rises to equal the value of pK_{a_2} (8.65), for all practical purposes, the α-COOH group will have ionized completely but the α-NH$_3^+$ group will also have deprotonated and one-half will exist as the uncharged NH$_2$ form. The desired *balance* point rests exactly midway between these extremes: $(1.94 + 8.65)/2$ or 5.30. At this pH value the A$^+$ species dominates since it represents the midpoint between the desired transition of the α-COOH group to COO$^-$ and the undesired loss of the charged α-NH$_3^+$ group in its conversion to NH$_2$. At a pH lower than 5.3, the loss of the charged α-NH$_3^+$ group is reduced but at the expense of COO$^-$ formation. At a higher pH than 5.3, COO$^-$ formation is encouraged but at the sacrifice of NH$_3^+$.

Finally, let us analyze the matter of the actual concentration of isoelectric ornithine (A^0) at pH 7.8. Since the α-COOH group has a pK_a value of 1.94, the A^{2+} species occurs in such an infinitesimally small amount that it can be ignored. Similarly, the anionic species (A$^-$) can also be ignored since it comprises approximately 0.1% of the total population. Consequently, only the cationic (A$^+$) and isoelectric (A^0) form need to be considered. The transition of A$^+$ to A^0 is governed by pK_{a_2} which has a value of 8.65. Thus

$$pH = pK_{a_2} + \log \frac{[A^0]}{[A^+]}$$

$$7.80 = 8.65 + \log \frac{[A^0]}{[A^+]}$$

$$\log \frac{[A^0]}{[A^+]} = -0.85 \text{ or } \frac{[A^+]}{[A^0]} = \frac{7.08}{1}$$

Since the ratio of A^+ to A^0 is 7.08 to 1.0, 1/8.08 or approximately 12.4% of the total ornithine population exists at pH 7.80 as the isoelectric species. The remaining 87.6% constitutes the cationic species A^+. While the above percentages are extremely close to the actual values, they deviate marginally since a minute amount of A^{2+} and A^- exist at pH 7.8.

As the pH increases, an ever-greater proportion of A^0 is produced; it reaches 50% at pH 8.65. Above 8.65, ornithine deprotonation comes to be governed increasingly by the transition of A^0 to A^-. At pH 9.65, the ratio of A^0 to A^+ is 10 to 1:

$$pH = pK_{a_2} + \log \frac{[A^0]}{[A^+]}$$

$$9.65 = 8.65 + \log \frac{[A^0]}{[A^+]}$$

$$\log \frac{[A^0]}{[A^+]} = 1.0 \text{ or } \frac{[A^0]}{[A^+]} = \frac{10.0}{1.0}$$

The contribution of A^- cannot be ignored since a significant amount of the anionic species exists at pH 9.65. The ratio of A^0 to A^-, governed by pK_{a_3} having the value of 10.76, is 12.88 to 1.0. It is determined as follows:

$$pH = pK_{a_3} + \log \frac{[A^-]}{[A^0]}$$

$$9.65 = 10.76 + \log \frac{[A^-]}{[A^0]}$$

$$\log \frac{[A^-]}{[A^0]} = -1.11 \text{ or } \frac{[A^0]}{[A^-]} = \frac{12.88}{1}$$

The above ratios must be expressed relative to a common factor, e.g., the isoelectric species. If there is 1 part A^+, then for every 10 parts of A^0, there is 0.776 part A^-. Recall that the ratio of A^0 to $A^- = 12.88$ to 1.0. As a result, the ratio $A^+:A^0:A^-$ is equal to 1:10:0.78. The amount of each species can be determined simply by taking each in relationship to the total number of parts. $A^+ = 1/11.78$ or 8.5%, $A^0 = 10/11.78$ or 84.9%, and $A^- = 0.78/11.78$ or 6.6%. At the higher pH of 9.65 as compared to 7.8, it is evident that significant error would have resulted had the contribution of the anionic species been ignored.

It is worthwhile to note that the pK_a of the carboxyl group of an aliphatic amino acid such as glycine is about 2.3 while that of a structurally similar fatty acid, e.g., acetic acid, at 4.8 is much higher. The tendency of the carboxyl group of the amino acid to deprotonate is over 300 times greater than that of the corresponding aliphatic acid. This results from the presence of a positively charged amino group, for the electron-withdrawing property of this group greatly enhances repulsion of the labile proton of the acid function. The ability of the amino group to affect carboxyl ionization would be expected to diminish as the amino group moves further from the α-carbon. This is confirmed by comparing the pK_a for β-aminopropionic acid of 3.6 with that of 4.2 for γ-aminobutyric acid. For the same reason, the proton of the amino group is held more tenaciously; it becomes a stronger base than a corresponding aliphatic amine and has a higher pK_a value.

It would be inappropriate to end this consideration of amino acids and their optical and dipolar properties without some mention of an interesting historical application of this information (see Greenstein and Winitz, 1961). In 1858, Louis Pasteur made the classical observation that the ascomycete *Penicillium glaucum* proliferates on racemic ammonium tartrate but only by the preferential consumption of the dextrorotatory form. This observation led Schulze and Bosshard, some 30 years later, to cultivate this microorganism on racemic glutamic acid or leucine. The specific optical rotational values of the unmetabolized compounds were opposite in direction but of equivalent magnitude to reference compounds secured from fungal protein hydrolysates. In one bold stroke, these workers employed a living system to resolve a racemic mixture and provided the first suggestion that organisms use metabolites of a stereomeric type akin to their proteins. This was the first experimental evidence for the L-directed stereospecific action of enzymes, although this level of significance could not be appreciated at that time. By the opening of the new century, however, gram quantities of racemic amino acids were being fed to animals for the selective excretion of the D-isomerides. In this way, quite pure D-stereomers of both protein and non-protein amino acids were obtained.

G. SPECIFIC OPTICAL ROTATION

As mentioned previously, naturally occurring amino acids characteristically are optically active compounds. Since the degree to which the plane of plane-polarized light is rotated is a function of the wavelength of the light, it has become customary to utilize the sodium D line at 589 nm as the monochromatic light source for measuring optical rotation. The actual rotation (α) is proportional to the concentration (c) of the amino acid and the length of the light path (l).

$$\alpha \propto cl$$

The constant of proportionality, k, is called the "specific optical rotation" and is designated by [α]. For an amino acid solution,

$$k = [\alpha]_\lambda^t = \frac{\text{observed rotation (°)}}{\text{length of sample tube (dm)} \times \text{substance concentration (g/100 ml)}}$$

The observed optical rotation is reported as specific optical rotation at a given temperature (t) and a particular wavelength (λ). The exact notation of these two parameters is most important since the observed rotation is affected markedly by the wavelength of the beam. Temperature is not such a critical component since, over the range of 20°–40°C, optical rotation is rarely altered more than 0.3° per degree rise in temperature, but wider variation can occur (Greenstein and Winitz, 1961).

$$[\alpha]_D^{20} = +26.5° \ (c, \ 0.014 \ \text{g/ml} \ H_2O)$$

The above notation specifies that the indicated optical rotation of +26.5° was determined at 20°C utilizing the sodium D line when 1.40 g of amino acid were dissolved in 100 ml of water. It is customary to utilize a sample tube of 1 dm and to prepare the sample at a 0.5 to 2.0% (w/v) concentration. The [α]$_\lambda^t$ values are generally taken in H_2O but other solvents including mineral and organic acids are used widely.

It is often desirable to express the optical rotation in the context of the compound's molecular weight. This is achieved by the use of molecular rotation, designated [M] and defined as

$$[M]_\lambda^t = \frac{[\alpha]_\lambda^t \times \text{molecular weight}}{100}$$

Prior to the availability of stereospecific enzymatic procedures, an amino acid's optical rotational properties were used traditionally as a criterion for determining its correct configurational family relationship. This point is exemplified in the Clough–Lutz–Jirgensons rule which states that the molecular rotation of an optically active α-amino acid of the L-configuration is shifted towards a more positive direction when placed under acidic conditions. A negative shift in rotation, however, is characteristic of a D-amino acid.

H. ABSOLUTE CONFIGURATION

The absolute configuration of the chiral center generated by the α-carbon atom of the amino acid can be determined by chemical or enzymatic procedures. Typical of a chemical method is the reaction of the α-amino acid with 2-methoxy-2,4-diphenyl-3(2H)-furanone to form pyrrolinone-type chromophores having an absorption maxima at 370 to 390 nm (see Toome and Reymond, 1975).

Gas chromatographic techniques have also been used in resolving configurational questions. In one application, the enantiomers are separated by an optically active supporting phase (see Parr and Howard, 1973). Another approach relies upon a chiral reagent to resolve the enantiomers (see Pollock *et al.*, 1965). On occasion, derivatives of the optical antipodes can be isolated directly; for example, N-trifluoroacetyl methyl esters since the L-isomer emerges before the D-isomeride (Hasegawa and Matsubara, 1975).

Among the best-known enzymatic approaches is the use of stereospecific enzymes such as the amino acid oxidases. D-Amino acid oxidase, a flavoprotein customarily obtained from hog kidney, specifically oxidizes D-amino acids to their corresponding oxo derivative, NH_3, and H_2O_2. Hydrogen peroxide, a strong oxidizing agent, must be degraded enzymatically with catalase to avoid decarboxylation of the oxo acid.

$$R-CH(NH_2)COOH + H_2O + O_2 \rightarrow R-C(=O)COOH + NH_3 + H_2O_2$$

$$R-C(=O)COOH + H_2O_2 \rightarrow R-COOH + CO_2 + H_2O$$

Although all D-amino acids are potential substrates for this enzyme some exhibit only limited activity.

L-Amino acid oxidases also occur in nature; the mammalian enzyme has only limited catalytic activity but it is a potent component of the venom of certain snakes. This enzyme has a remarkable degree of stereospecificity and it also produces the corresponding oxo derivative NH_3 and H_2O_2 from a given L-amino acid. A large molar excess of the D-amino acid does not impede oxidation of the L-amino acid. This makes the enzyme suitable for dealing with contamination of the D-isomeride by trace quantities of the L antipode. Thus, L- and D-amino acid oxidases, available commercially, are not only used for configurational determination but also to remove the unwanted D- or L-isomer from a racemic mixture (see Hardy, 1974).

Another stereospecific enzyme application involves the acetylation of the D-amino acid. Baker's yeast contains an N^α-acetyltransferase which mediates, with a very high degree of specificity, transfer of the acetyl group to form the N^α-acetyl derivative of the amino acid. Acetyl-CoA is the physiological donor. Reaction of this enzyme with a racemic mixture of amino acids and labeled acetate produces radioactive acetylated amino acids only for the D-amino acids. While this enzyme lacks reactivity toward proline and exhibits only a sparse reaction rate with the dicarboxylic acid and hydrophobic amino acids, it does act broadly and exclusively with D-amino acids (Schmitt and Zenk, 1968).

Finally, mammalian kidney is a prolific producer of an enzyme, known trivially as acylase I, that possesses marked specificity toward N-acylated derivatives of α-amino acids of the L configurational family, particularly when an aliphatic chain is present. A racemic amino acid mixture can be acylated chemically prior to reaction with acylase I. This procedure will generate from the mixture only the free amino acid of the L-isomer since the D-amino acid retains its N-acylated group.

ADDITIONAL READING

Greenstein, J. P., and Winitz, M. (1961). "Chemistry of the Amino Acids," 3 vols. Wiley, New York.

IUPAC Committee on the Nomenclature of Organic Chemistry and IUPAC-IUB Committee on Biochemistry Nomenclature (1975). Nomenclature of α-amino acids. *Biochemistry* **14,** 449–462.

Jakubke, H.-D., and Jeschkeit, H. (1977). "Amino Acids, Peptides and Proteins—An Introduction." Wiley, New York.

Kopple, K. D. (1966). "Peptides and Amino Acids." Benjamin, New York.

Larsen, P. O. (1980). Physical and chemical properties of amino acids. *In* "The Biochemistry of Plants" (B. J. Miflin, ed.), Vol. 5, pp. 225–269. Academic Press, New York.

Nivard, R. J. F., and Tesser, G. I. (1965). General chemistry of the amino acids. *Compr. Biochem.* **6,** 143–207.

O'Connor, R. (1974). "Fundamentals of Chemistry—A Learning Systems Approach." Harper & Row, New York.

Segel, I. H. (1968). "Biochemical Calculations," 2nd ed., Wiley, New York.

Chapter 2

Analytical Methodology

In a simple set of operations and at one dramatic moment the entire complement of ninhydrin-reactive nitrogen compounds could be seen dispersed upon
the paper sheet. The obvious amino acids and the two amides were immediately recognized, and their relative amounts were surmised from the
intensity of the ninhydrin color by which they were revealed. Even such amino
acids as β-alanine, readily recognizable by its blue color with ninhydrin, and
threonine, previously recognized only very rarely in plants, could be seen.
However, one conspicuous ninhydrin-reactive spot occurred directly below
valine on phenol-collidine-lutidine chromatograms. This unexpected substance
proved to be γ-aminobutyric acid. . . .

Steward and Durzan (1965) commenting on one of the earliest applications of
two-directional paper chromatography for the analysis of a plant extract
(potato) by Dent *et al.* (1947).

A. INTRODUCTION

The revolutionary advancement in the isolation and purification of
non-protein amino acids which has occurred during the past third of a
century has swelled the ranks of known structures and compounds.
This expanded inventory is due largely to the application of paper partition and ion-exchange chromatography. The chromatographic separation of amino acids dates from 1941 when Martin and Synge effected
their classical separation of N-acetylated amino acids utilizing a silica gel
column; chloroform served as the developing solvent. Three years later,
Consden, Gordon, and Martin combined the ease of chromatographic
separation on a sheet of filter paper with the principles of liquid partitioning to pioneer two-dimensional paper partition chromatography.
The work of Dent *et al.* (1947) deserves mention since it was among the

earliest applications of partition chromatography with filter paper sheets to the detailed analysis of amino acids and other ninhydrin-positive substances.

Moore and Stein (1949) experimented with a separational process that employed columns containing starch, but it was abandoned for a superior liquid separational process with ion-exchange resin (Dowex-50) which they advanced in 1951. By 1958, their ion-exchange chromatographic methods were automated (Spackman *et al.*, 1958). Over the past 20 years, several alternative analytical systems have appeared, such as gas–liquid chromatography, but at this time none have replaced significantly the automated procedures of Moore and Stein and their collaborators which presently dominate amino acid analysis.

B. PARTITION PAPER CHROMATOGRAPHY

When a given compound is added to a mixture of partially immiscible solvents, the solute distributes according to its affinity for each solvent. The ratio of the solute concentration at equilibrium in the two phases is known as the *partition coefficient*. All amino acids exhibit an experimentally determinable partition coefficient for a given immiscible solvent system.

As part of their classical study of the amino acid constituents of wool, Martin and Synge (1941) attempted to separate the *N*-acetylated derivatives by partitioning in an extractor apparatus containing 40 mixing vessels; chloroform and water served as the opposing immiscible solvents. The poor resolving capacity of this procedure was immediately evident since it reflected the low intrinsic efficiency of this technique. These workers correctly reasoned that if one of the liquid phases could be rendered immobile while the second phase was permitted to flow over the immobile phase, a given amino acid would have a very large interface in which to attain an equilibrium distribution between the phases. To achieve this goal, a chromatographic column was filled with silica gel which retained the aqueous solvent so tenaciously as to be rendered immobile. The column was "developed" with mobile chloroform.

Roughly speaking, partition paper chromatography as conceived by Consden *et al.* (1944) relies on the migration of a given solvent system through an immobilized cellulose–water complex. Paper chromatographic sheets consist of a fibrous network of refined cellulose molecules capable of sequestering appreciable water amounting to 20–25% of the dry paper's weight. These molecules are organized first into microfibrils and eventually macrofibrils; the latter are oriented into regions of highly amorphous zones. It is believed that the cellulose molecules, situated in these amorphous zones, interact with water. Various components in the sample exhibit differential affinity for the stationary cellulose–water phase and the migrating solvent system and eventually they attain what may be approximated as an equilibrium distribution between these phases. A given amino acid partitions between the stationary phase created by the paper–water complex and the mobile solvent system—hence the term paper partition chromatography.

The migration of a particular amino acid closely approximates but does not comply with that predicted for simple liquid–liquid partitioning. This deviation reflects the fact that paper partition chromatography is strictly speaking not a true partition chromatography since the supporting phase (the paper) is not entirely inert. The presence of carboxyl groups impart a weak but discernible ion-exchange property to the paper [a factor also shown by the strong affinity of basic amino acids for cellulose (see Spener and Dieckhoff, 1973)]. In addition, while the stationary phase can attain a balance with the vapors of the mobile phase, a true equilibrium is usually not achieved between these phases. Variation ensues in the absolute amount and ratio of the phases along their common interface; this causes an amino acid migration pattern that deviates somewhat from that predicted by partition theory.

Solute migration is variable and affected by a host of parameters including pH; temperature; paper geometry, constitution, and orientation; distance of the solute from the solvent reservoir; solvent quantity; solute quantity; extraneous substances and the duration of the run. Of the many variable parameters, none seem more significant than temperature and pH. With certain solvent systems the influence of temperature is minimal, e.g., phenol-NH_3, but most solvent systems are subject to much greater temperature-dependent perturbation.

Alteration in solvent pH has a pronounced effect on the resulting R_f value of the chromatographed amino acid (Fig. 2). The pH effect is so pervasive that even the pH of the sample prior to chromatography can alter the ultimate sample migration and thereby affect the R_f values and overall resolution. The influence of pH on R_f values is dramatically revealed by the data of Fig. 3 taken from the research of Landau *et al.* (1951). Of the two solvent systems shown in Fig. 3, phenol is much more susceptible to such R_f value fluctuations than is 2,4-lutidine or n-butanol (the latter is not shown in Fig. 3). Each amino acid forms a maximally compacted spot at some particular pH value or range and it may even be possible to utilize this factor to aid in resolving certain compounds by manipulating the pH of the sample.

Two-dimensional paper chromatographic resolution of nonprotein amino acids can be achieved with many solvent system components

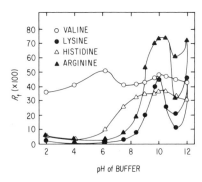

Fig. 2. Variation in amino acid R_f values as a function of the developing buffer pH. The solvent system used was *m*-cresol. The ordinate axis is reproduced accurately but should have been labeled $R_f \times 100$. Reprinted with permission from McFarren (1951). Copyright by the American Chemical Society.

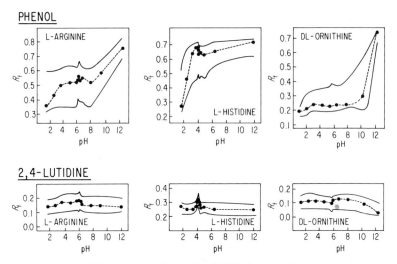

Fig. 3. Amino acid R_f values and sample pH. The lower and upper curves represent the bottom and top, respectively, of the detected spot, while the point denotes the center of the spot produced by the indicated sample pH. The measured R_f values are presented on the ordinate. Reprinted with permission from Landau *et al.* (1951). Copyright by the American Chemical Society.

including phenol, *m*-cresol, *n*-butanol, *tert*-butanol, collidine, lutidine, and methyl ethyl ketone. Claims for the intrinsic superiority of one system over another are commonplace and these should be viewed with caution. Rockland and Underwood (1954) obtained a fine separation with *tert*-butanol:water:formic acid (69.5:29.5:1, v/v) followed by phenol-NH_3. They contended that the initial solvent system represented a superior liquid phase since it avoided the disagreeable odor and high cost of collidine–lutidine mixtures and variations in solvent composition due to differential esterification of *n*-butanol in alcohol–acid mixtures. At the same time, it proved to have a greater resolving capacity than the other tested solvents. Phenol and *m*-cresol were found to be more effective when combined as compared to their separate use. As a result, Levy and Chung (1953) employed *n*-butanol:acetic acid:water (4:1:5, v/v) prior to *m*-cresol:phenol (1:1, w/w) in pH 9.3 borate buffer to produce their high quality chromatographs.

The components of a diheme peptide isolated from the photo-anaerobic bacterium *Chromatium* were fully separated by a mixture of *n*-butanol:methyl ethyl ketone:water:28% NH_3 (25:15:7:3, v/v) run in the first dimension for 40 to 45 hr followed by *n*-butanol:acetic acid:water (4:1:5, v/v) for 22 to 24 hr (Dus *et al.*, 1962). These authors reported that their solvent system, as compared to phenol, provided greater resolution, ease of handling, sharper spots, and a lack of background coloration upon spraying. Sample loads of 50 to 100 nmoles were optimal; the resulting distributional pattern constitutes Fig. 4A. Several interesting generalizations on the relationship of amino acid structure to their migrational pattern resulted from the work of Dus *et al.* (Fig. 4B) and their observations are presented directly:

> In accordance with expectation based on this diagram (Fig. 4B) diaminobutyric acid should not move as fast in the second direction as do the monoaminobutyric acids but, rather, should migrate as lysine does. Also diaminopimelic acid should be strongly retarded in the second direction, compared to the corresponding monoamino acids. As the number of carbon atoms in the side chains increases, the amino acids should move faster in both directions, so that in the series (glycine, alanine, valine, leucine) all of the amino acids should lie on a diagonal. If the amino group is shifted from the α-position to the β, γ, δ, or ϵ positions, increasing retardation is observed in the first direction, so that α-aminobutyric acid moves faster than the β form, and the β form faster than the γ-form, and so forth. The placement of a hydroxyl group in a side chain has a remarkable effect, namely, that the hydroxylated amino acids tend to concentrate ˙on or near a narrow region forming a line across the diagram; i.e., serine, homoserine, threonine, and allothreonine are exactly in line, and hydroxyproline and hydroxylysine are almost in line. Tyrosine, although not in line, is strongly retarded in the second direction in comparison with phenylalanine. With these trends in mind, it is possible, when observing an unknown amino acid spot on the chromatogram, to make an intelligent guess as to the structure of the unknown amino acid.

Among the multitude of solvent systems developed for nonprotein amino acid analysis, undoubtedly the most popular consists of aqueous phenol to which NH_3 is added; it is followed by *n*-butanol:acetic acid:water (4:1:5,v/v). The use of these solvents to effectively separate

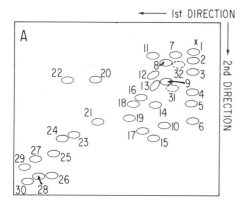

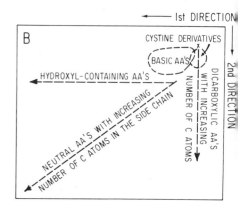

Fig. 4.   (A) Amino acid distribution obtained by a two-dimensional chromatographic procedure. The solvent system consisted of *n*-butanol–methyl ethyl ketone–water–concentrated ammonia (25:15:7:3, v/v) and was used in the first direction for 45 hr. *n*-Butanol–glacial acetic acid–water (4:1:5, v/v), 1 day old, was employed in the second direction for 22 hr. 1, Cysteic acid; 2, homocysteic acid, 3, aspartic acid; 4, glutamic acid; 5, α-aminoadipic acid; 6, α-aminopimelic acid; 7, lysine; 8, arginine; 9, glycine; 10, γ-aminobutyric acid; 11, histidine; 12, serine; 13, methionine sulfone; 14, β-alanine; 15, δ-aminovaleric acid; 16, alanine; 17, β-aminobutyric acid; 18, proline; 19, α-aminobutyric acid; 20, allothreonine; 21, tyrosine; 22, threonine; 23, valine; 24, methionine; 25, ethionine; 26, norleucine; 27, tryptophan; 28, iso(+)-alloisoleucine; 29, phenylalanine; 30, leucine; 31, hydroxyproline; and 32, hydroxylysine. (B) Distributional pattern of the amino acids resolved in the above solvent systems. Reprinted with permission from Dus *et al.* (1962). Copyright by the American Society of Biological Chemists, Inc.

nonprotein amino acids is revealed by the developed chromatogram illustrated in Fig. 5.

By a selection of appropriate solvent systems and operating conditions, a very high degree of resolution can be attained routinely even with complex amino acid mixtures. For example, Gray *et al.* (1964) achieved the separation of the diastereomeric forms of several amino acids and effectively resolved isoleucine, leucine, and norleucine as well as their higher homolog. The final selection of the particular solvent system and operating conditions are usually derived empirically but some guiding principles have been established. Several examples enunciated by Greenstein and Winitz (1961) are

(a) for a homologous series of solvents, the R_f value of a solute will vary inversely as the molecular weight of the solvent; (b) the R_f value of a solute tends to increase with additional water content of the solvent system (this effect can be achieved by mixing a solvent with a polar liquid prior to saturation with water, i.e., lutidine with collidine); (c) if the substance is a weak acid or base, depression of ionization will tend to increase its R_f and vice versa. Thus, the addition of ammonia to a phenol system retards the dicarboxylic amino acids and accelerates the movement of basic amino acids.

Partition paper chromatography has responded efficaciously to preparative needs as well. Generally, thick paper is employed, e.g., Whatman No. 3 or 3 MM and up to 25 mg of sample are streaked across one length of the paper at the origin. The location of the desired compound is determined by cutting a narrow strip from the center and sides of the chromatogram and treating these samples with an appropriate detecting reagent. While this process is still used occasionally, modern advances in ion-exchange chromatography have largely made this procedure obsolete.

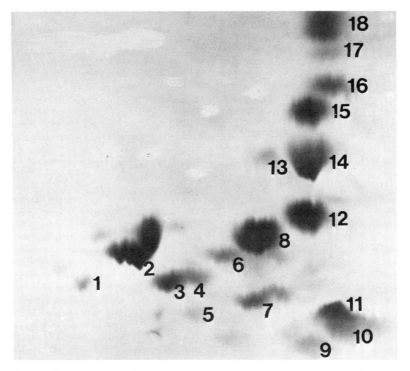

Fig. 5. Two-dimensional partition paper chromatogram of a seed extract of *Aesculus parviflora*. 1, Aspartic acid; 2, glutamic acid; 3, serine; 4, glycine; 5, asparagine; 6, threonine; 7, glutamine; 8, alanine; 9, histidine; 10, lysine; 11, arginine; 12, (*exo*)-*cis*-3,4-methanoproline; 13, tyrosine; 14, γ-aminobutyric acid; 15, valine; 16, ethyl glutamate; 17, phenylalanine; and 18, leucine. Photograph kindly furnished by L. Fowden.

We have seen that the R_f value of different runs can vary. If, however, ambient conditions are restricted to a narrow operating range, the high quality and purity of present day reagents and paper permit such remarkably consistent runs within a given laboratory that after a series of analyses, one quickly recognizes the constituent amino acids simply by their final position on the chromatogram and the juxtapositioning of the detected spots. When a compound possessing unique R_f values appears on the chromatogram, it is evident immediately by virtue of the altered visual pattern of the ninhydrin-reactive spots. If chromatograms are prepared under variable ambient conditions, the chromatographic process can be rendered more reproducible by relating sample component migration to an arbitrarily selected reference amino acid rather than the transit of the solvent front.

Chromatographic runs with several sheets of paper are advantageous since they permit simultaneous analyses of reference standards under the same operating conditions. This facilitates verification or compilation of the R_f values for known substances of interest. Similarly, the addition of radioactive compounds to the sample mixture allows the identification of the putative spot to be confirmed by autoradiography or another suitable detection method. Whatever the means of verification, it is most important not to rely on the coincident migration of a sample with a reference standard as the sole criterion for establishing the identity of a particular amino acid (see Rosenthal, 1978c). This caution is not limited to partition paper chromatography but is applicable to all methods described in this chapter.

C. THIN-LAYER CHROMATOGRAPHY

Numerous supporting media other than filter paper sheets can be employed for partition chromatography and electrophoresis. Silica gel, microcrystalline and other cellulosic preparations, aluminum oxide, Kieselguhr G, magnesium silicate, and even activated charcoals have functioned as the supported stationary phase. The above materials are spread as a finely granulated or *thin layer* over a solid support such as glass; a binding agent, e.g., gypsum or starch may be included. In this procedure, reproducible chromatograms are achieved by making the adsorptive layer as uniform as possible. Use of thin-layer chromatograms failed to enjoy widespread acceptance until the development by Stahl of a mechanical device that could rapidly, easily, and reproducibly spread 150 to 250 μm-thick coatings of a slurry of the stationary phase on a glass plate. Stahl contributed significantly to the development of

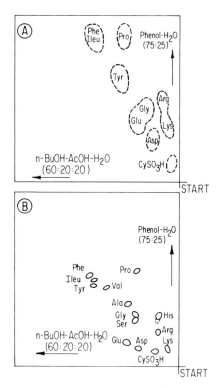

Fig. 6. Comparison of paper partition and thin-layer chromatography. (A) Whatman no. 1 paper; (B) silica gel G. Reprinted with permission from Brenner and Neiderwieser (1960). Copyright by the Birkhaüser Verlag.

modern thin-layer chromatography, not only by pioneering many procedural developments but also by publicizing adroitly the utility and advantageous characteristics of this analytical method.

Thin-layer chromatography has the distinctive benefit of lacking a solid phase made of paper which has a fibrous structure; this avoids sample spreading along fiber boundaries. The shorter "run" time associated with this procedure minimizes diffusion of the sample spot and lowers the limit of detection of these more concentrated spots. As a result, the separations can be sharper and more fully resolved while necessitating less sample than in paper partition chromatography* (Fig. 6).

---

*On the other hand, as inspection of Fig. 5 reveals, excellent results are obtainable by paper partition chromatography.

Thin-layer plates are characteristically 20 × 20 cm in configuration but they can be cut into 5 × 20 cm strips if a single dimensional chromatographic run is desired. Thin-paper plates are developed in specially designed glass chromatographic tanks that conserve solvent and have the added advantage of rapid equilibrium between the solvent system and the chamber vapor phase.

Preparative procedures merely require a thicker solid support but this applies only to a certain point. As a generalization, with silica gel G, an average of 1 mg of material can be processed per cm of chromatogram width per mm of solid support thickness. The solvent systems are the same as those used in paper partition chromatography. A combination of partition chromatography and electrophoresis can be employed with thin-layer plates and this procedure is often the method of choice. Thin-layer chromatography is valued primarily for the rapidity of analysis, its sharpness, marked sensitivity, and the greater flexibility afforded in the use of various stationary phases instead of relying solely on cellulose.

D. NINHYDRIN COLOR REACTION

Amino acids in solution are colorless and their common functional groups lack convenient absorption maxima since they exhibit little absorption of ultraviolet radiation above 200 nm. Other than aromatic and heterocyclic exceptions, nonprotein amino acids characteristically lack side chain chromophores exhibiting absorbance in the UV region of the spectrum. As a result of these factors, their detection relies generally upon color-producing chemical reactions. The most useful technique is that of simply spraying the amino acid-laden material with a solution of ninhydrin, such as 0.25% (w/v) ninhydrin in acetone, and then heating at 100°C for 5 to 10 min. The use of ninhydrin for detecting amino acids on a paper chromatogram permits visualization of around 1 to 3 nmoles of such compounds as glycine and histidine.

The ninhydrin procedure is predicated upon the reaction of ninhydrin (triketohydrindene hydrate) (I) with an α-amino acid (II) to form a variously colored complex which, while it is known as Ruhemann's purple (VII), is characteristically shades of blue, purple, or violet. The initial reaction was taken by Greenstein and Winitz (1961) to involve the oxidative deamination of the amino acid and proceeds through an amino acid intermediate (IV) to produce an α-keto acid (V) and NH_3; the corresponding aldehyde (VIII) and CO_2 are the eventual reaction products. Ninhydrin is reduced concurrently to hydrindantin (III) and then aminated to produce the colored component (VII). Several alternative proposals on the chemical basis for the ninhydrin reaction have appeared. The most detailed being that of McCaldin (1960; see also Lamothe and McCormick, 1973).

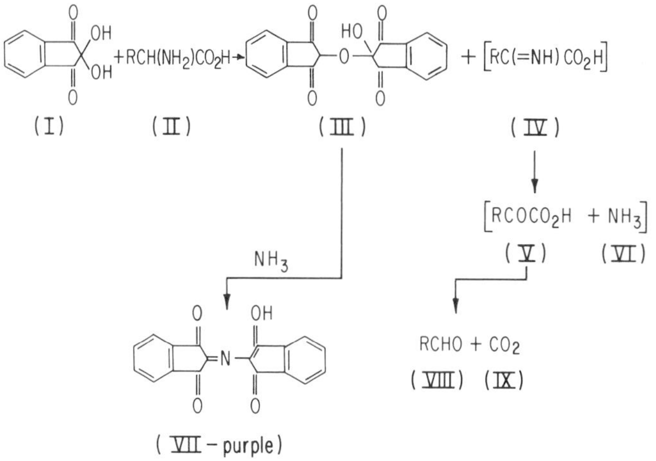

After Greenstein and Winitz (1961).

Ruhemann's purple, produced with various amino acids, has an absorbance maximum centering around 570 nm and this wavelength has become the accepted standard for ninhydrin analysis of amino acids. Although this chromogen forms in substoichiometric yields that vary somewhat for various amino acids, reproducible color is obtainable if a suitable reducing system blocks oxidative side reactions that can deplete the color. This reduction can be achieved with titanous or stannous chloride; other suitable reducing agents include cyanide and hydrazine.

In the reaction of certain imino acids (II) with ninhydrin (I), CO_2 is released without evolution of ammonia. The chromogen produced with imino acids is yellow (III) and possesses a broad spectrum centered around 440 nm.

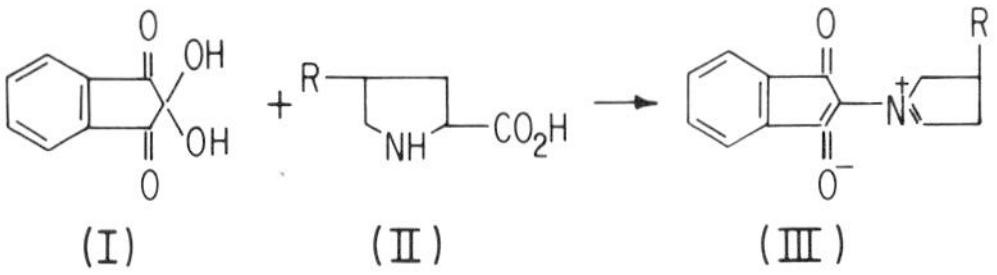

After Greenstein and Winitz (1961).

Ninhydrin reacts with compounds other than α-amino acids, such as primary and secondary amines (not tertiary or aromatic amines), amino alcohols, and even ammonia and all ammonium salts. Ninhydrin can also form other color complexes with specific amino acids, e.g., red with lathyrine, green with β-cyanoalanine, while γ-methyleneglutamic acid and azetidine-2-carboxylic acid give brown, histidine forms red-gray, and 5-hydroxypipecolic acid forms blue. Atypical color formation is not a disadvantage of this reagent since these striking colors can play a decisive role in the initial discovery and subsequent isolation of a novel nonprotein amino acid.

Several modifications of the basic ninhydrin spray reagent have been described; particularly useful ones involve the addition of metallic ions such as cadmium or copper or organic solvents (e.g., ethanolamine) that impart distinctive coloration to the heated amino acids. The $Cu(NO_3)_2$-ninhydrin reagent developed by Moffat and Lytle (1959) is probably the best known polychromatic reagent for amino acid detection. A solution containing 0.2% (w/v) ninhydrin in absolute ethanol:glacial acetic acid:2,4,6-collidine (50:10:2, v/v) is mixed with 1% (w/v) ethanolic $Cu(NO_3)_2 \cdot 3H_2O$ so that 25 parts of the first solution is present per 1.5 parts of the ethanolic cupric nitrate. The distinctive colors are developed by heating at 105°C for 2 min. Selective color formation varies with several tested solvent systems, with n-butanol:acetic acid:water (4:1:5, v/v) and butanol:water (15:85, v/v) being the most effective in eliciting selective color production (Moffat and Lytle, 1959).

Another polychromatic detection system involves exposing the chromatogram for 2 min to diethylamine at room temperatures. (The vapors are generated by evaporating the reagent from an open dish placed at the base of the chromatography chamber). The diethylamine-treated chromatograms are then sprayed with ninhydrin in the usual manner. These reagents are as sensitive as ninhydrin itself and claimed by the authors to be superior to cupric nitrate–ninhydrin for differentiating between certain amino acids (Circo and Freeman, 1963). These polychromatic reagents have not been applied systematically to nonprotein amino acid identification where they may well be of real diagnostic value.

Kawerau and Wieland (1951) have provided a means for preserving the distributional pattern of the amino acids by spraying the ninhydrin-developed chromatogram with a solution consisting of 1 ml of saturated aqueous $Cu(NO_3)_2$ and 0.2 ml of 10% (v/v) HNO_3 taken to 100 ml with 95% aqueous ethanol. This spray forms a light-stable, red, copper-containing pigment; since the pigment is acid labile, the chromatogram should be thoroughly exposed to ammonia vapors after spraying. For long-term storage, these authors recommended dipping the paper in a saturated solution of "Perspex" (methyl methacrylate polymer) in chloroform.

Other interesting modifications of the ninhydrin reagent are provided in the handbook "Data for Biochemical Research" edited by Dawson *et al.* (1969). This manual provides a wealth of information on reagents that produce specific colors with particular functional groups of amino acids as well as with various derivatized forms of these compounds.

E. ION-EXCHANGE CHROMATOGRAPHY

1. Ion-Exchange Resin

Ion-exchange chromatography is a separation process based on the use of insoluble polymeric materials carrying exchangeable cations or anions. It is a very powerful and efficacious means for molecular separation and its utility in the isolation, purification, and identification of amino acids is particularly noteworthy. A typical ion-exchange resin can be prepared by copolymerizing styrene and divinylbenzene to produce a polymeric matrix stabilized by periodic cross-linking groups. The cross-linked polystyrene is then sulfonated to generate a strongly acidic exchange resin possessing charged SO_3^- groups as an integral part of the three-dimensional polymeric structure of the resin. This charged group attracts ions of the opposite charge (cations) which become distributed within the free space of the resin. These attracted cations are held electrostatically to the resin and can be exchanged for an equivalent quantity of another similarly charged ion. Thus, the incorporated SO_3^- group functions in the exchange of cations, hence the term cation-exchange resin. This particular resin is known commercially by several trade names including Dowex-50; it is one of the most commonly used cationic exchange resins for analytical and preparative procedures with amino acids.

Such sulfonated polystyrene resin is quite stable and it can be used for protracted periods even at 100°C. It behaves as a strong acid which results in a very constant exchange capacity at pH levels above 2. Commercial preparations of this type of resin routinely have an exchange capacity of 4 to 5 mEq/gm dry resin which represents nearly one sulfonic acid group per aromatic ring. A cationic exchange resin less acidic than Dowex-50 is produced when a weaker acid, such as an organic acid, is incorporated into the resin matrix. The availability of such carboxylic acid-containing resins can be most useful since they exchange only with strong or moderately strong bases.

On the other hand, a very basic ion-exchange resin is created by introducing a quaternary ammonium group into the resin; it retains its counter Cl^- at all but very alkaline pH values since its apparent pK value rests in excess of 13 (Dowex-1). Another highly basic resin (Dowex-2) is generated by incorporating a secondary amine into the resin but a far less basic anionic exchange resin, i.e., Dowex-3, with pK values much nearer to neutrality (pH 7 to 9), results from inclusion of a primary amine. The exchange capacity for such strongly basic resins as Dowex-1 or -2 is about 3 to 4 mEq/gm dry resin while Dowex-3 has a greater exchange capacity, on the order of 5 to 7 mEq/gm dry resin.

The chlorovinylbenzene concentration of the original polymer mix determines only the approximate degree of resin cross-linking, although it is commonly taken to represent the divinylbenzene content of the total polymer mix. This parameter is indicated by the designations X-1, X-2, X-4, etc., where the numerical value stands for the nominal divinyl-benzene content as a percent of the polymer mix. Commercial res-ins are manufactured with a divinylbenzene content of from 1 to 16% but 4 to 8% cross-linked resins are suitable for most amino acid applica-tions. Increased cross-linking of the resin diminishes resin swelling and enhances its overall ion-exchange capacity since the number of ex-changeable sites per unit volume of resin are increased. At the same time, the equilibrium time is extended, since the higher the degree of cross-linking, the less permeable the resin is to solute molecules. This factor can be exploited in the separation of amino acids from large mac-romolecules since the latter substances cannot readily penetrate highly cross-linked resins.

2. Ionic-Exchange Process

Williams and Wilson (1975) have characterized the actual ion-exchange mechanism in terms of five distinctive stages: (a) Rapid diffu-sion of the ion to the resin surface. (b) Diffusion from the resin surface to the exchange site. This stage is, of course, related to the solute concen-tration and the degree of cross-linking. (c) Instantaneous exchange of the labile cation or anion:

$$R_1—SO_3^- ......^+H \ + \ H_3^+N—R^1 \ \leftrightharpoons \ R_1—SO_3^- ......H_3N^+—R^1 \ + \ H^+$$

$$R_1—^+N(CH_3)_3 ......^-Cl \ + \ ^-OOC—R^1 \ \leftrightharpoons \ R_1—^+N(CH_3)_3 ......^-OOC—R^1 \ + \ Cl^-$$

(d) Diffusion of the exchanged ion to the resin surface. (e) Desorption by the effluent and diffusion of the exchanged ion into the external solu-tion.

In essence, ion exchange chromatography is a process that employs an insoluble polymeric substance containing charged groups as part of its polymeric structure. Balancing these charged groups are labile counter ions of opposite charge; the latter ions can be exchanged stoichiometrically for other ionic species (e.g., amino acids) of like charge. Since these processes are freely reversible, ion-exchange resins can be regenerated.

In the actual ion-exchange process that constitutes elution chromatography, the amino acid-containing sample is applied to a column of appropriately charged resin. As the sample percolates through the resin bed, none, some, or virtually all of the charged amino acids exchange with the resin's counter ion. This exchange process occurs repeatedly until the effective ion-exchange capacity of the resin is exhausted. Non-exchanged constituents are then washed fully from the resin bed with water or some other appropriate solvent.

To discharge the amino acids from the column, a "developing" solvent is passed through the column bed. The greater ion concentration in the developing solvent ultimately forces the bound amino acids from the resin as it is a concentration-dependent equilibrium process. If the developing solvent is selected properly, the bound substances will elute selectively so as to separate the "desired" components from contaminant substances or to release sequentially the components contained within the column (as in automated amino acid analysis).

In their discussions of "displacement chromatography," Partridge and Brimley (1952) asserted that the order of elution for monoamino-monocarboxylic acids parallels their dissociation constant for the reaction

$$A^+ \rightarrow A^\pm + H^+$$

This reflects the fact that the resin absorbs only cationic species (A^+) of the amino acid. In other words, as the amino acid's carboxyl group ionizes, the amino acid becomes more anionic, i.e., $A^\pm$ comes to dominate A^+. The amino acid has less affinity for the exchangeable resin sites with the increasingly greater concentration of cations in the developing solvent. Partridge and Brimley have shown that, with few exceptions, the pK_a of the amino acids predicts the elution profile of the "developed" amino acids. (Although their findings were developed for displacement chromatography, they are applicable to elution chromatography as well.)

Simply stated, in response to a gradient of increasing pH, acidic amino acids with their low pK_a values will elute first, followed by neutral members, and finally the basic components with their higher pK_a values. In fact, however, the elution profile not only reflects a gradient of increasing pK_a values but also results from interaction between the resin and the non-ionic portion of the amino acid. Thus, the attraction between the amino acid and the resin should be viewed as a composite of two factors: electrostatic attraction between appropriately charged species as well as van der Waals forces and hydrophobic interactions. The latter factors increase with greater molecular weight and are affected by structural features such as the presence of an aromatic ring. For example, as the carbon chain length of an aliphatic compound increases, the substance's attraction for nonpolar portions of the resin is enhanced and a greater column retention time results. An amino acid with a branched side chain elutes prior to one with an equivalently sized straight side chain, e.g., isoleucine and leucine before norleucine. The presence of a hydroxyl group accelerates compound elution, presumably due to enhanced affinity for the aqueous phase.

In developing a strategy for processing the amino acid fraction of a natural product extract, these materials can be isolated initially by use of a strongly acidic ion-exchange resin such as Dowex-50 in the H^+ form. This strongly acidic exchanger binds all zwitterions except such markedly acidic substances as cysteic acid. However, the protons displaced from the resin during the exchange process decrease the pH of the effluent and this can prevent certain acidic components from binding to the resin. Loss of these acidic components can be avoided by prior conversion of the resin to the pyridinium or 3-chloropyridinium form.

The acidic and neutral amino acids retained on Dowex-50 (H^+) can be eluted with aqueous ammonia; however, basic compounds, with their higher pK_a values, require a more concentrated ammonia solution before they are displaced from the resin. Concentration of the ammonia-laden effluent *in vacuo* yields neutral and basic amino acids in the free state while acidic members are secured either partly or fully as monoammonium salts. Inorganic cations are retained by the resin with ammonia-based development but these ions are eluted if the Dowex-50 resin is developed with HCl.

Neutral and acidic non-protein amino acids of a particular mixture can be segregated readily from basic components. Conversion of Dowex-50 to the NH_4^+ form prevents significant initial binding of neutral and acidic components. Basic compounds can then be obtained by development with ammonia (see below). This type of group separation can also be achieved by elution with aqueous pyridine regardless of whether a H^+ or pyridinium form of the resin is employed.

Acidic amino acids are retained tenaciously by such strongly basic resins as Dowex-1. In the OH^- form, this resin attracts all anions and zwitterions including weak and strong acids from solution. Weakly basic exchangers, on the other hand, bind only moderately strong or strong acids. With Dowex-1 in the acetate form, the neutral and basic constituents elute with the aqueous wash; retained acidic compounds can then be recovered by development with acetic acid.

3. Preparative Ion-Exchange Chromatography

In the case of nonprotein amino acids, ion-exchange chromatography is employed often in a preparative sense, i.e., for the isolation of a single component or groups of constituents, frequently in large amounts, from a particular biological source. These isolations are often a prerequisite to compound chemical characterization and biological study. An excellent example of the importance of selecting the appropriate resin and operating conditions is provided by examining the preparation of the moderately basic nonprotein amino acid L-canavanine (isoelectric point = 8.2) from a complex mixture of amino acids and other natural products of jack bean seeds. After obtaining canavanine and other soluble components by aqueous ethanolic extraction of several hundred grams of defatted seed meal, the extract is treated with Dowex-50 (NH_4^+). Unlike the H^+ form, the NH_4^+-containing resin has little affinity for acidic and neutral amino acids, and these compounds as well as uncharged species are removed extensively by thorough washing of the resin with water. Advantage is then taken of the moderately basic nature of canavanine and the resin is developed with a very dilute solution of NH_3. Under this condition, basic compounds such as arginine do not exchange appreciably with the NH_4^+ counter ion and are retained by the resin whereas canavanine is released. After the NH_3 is removed by evaporation *in vacuo* and the effluent decolorized with charcoal, canavanine of at least 96% purity is obtained routinely by a single crystallization. A second crystallization yields gram quantities of the nonprotein amino acid with a purity in excess of 99% (Rosenthal, 1977b).

It is not customary, however, to achieve such an efficacious purification in essentially a single discrete step. A more typical situation is illustrated in the procedures developed by Fowden *et al.* (1969) for the isolation from *Aesculus parviflora* of *cis*-α-(carboxycyclopropyl)glycine and *exo(cis)*-3,4-methanoproline. The aqueous ethanolic extract of the *Aesculus* seed was applied initially to a column of Zeokarb 225 (sulfonated polystyrene resin) in the H^+ form. Since cationic substances are sequestered by this resin, simple washing with water effectively removed many non-amino acid components. A 0.2 N NH_3 solution eluted both of the desired amino acids in the first 35 fractions along with such contaminants as acidic and hydroxy amino acids as well as proline. In this procedure, a relatively dilute ammonia solution was sufficient to rapidly remove the desired amino acids while unwanted components were retained. A more dilute solution might sacrifice the yield in the early fractions while not necessarily improving the purity of the preparation. A more concentrated solution might enhance the yield, but the potential for much greater contamination of the desired products cannot be overlooked.

The appropriate fractions from the first column were pooled, concentrated, adjusted to pH 7, and chromatographed on Dowex-1, a strongly basic quaternary resin. *exo*(*cis*)-3,4-Methanoproline and several known contaminants, too basic to be held to the column, were eluted with the aqueous wash. They were processed further on a Dowex-50 column (H^+ form), eluted with 0.2 N NH_3, and crystallized from ethanol–acetone–water. On the other hand, *cis*-α-(carboxycyclopropyl)glycine exchanged onto the Dowex-1 column and was eluted subsequently with 0.3 N acetic acid. Further purification of this exocyclic compound was achieved by partition paper chromatography of the Dowex-1 effluent.

An unusual aspect of this purification procedure for *exo*(*cis*)-3,4-methanoproline is the repetitive treatment with Dowex-50 (H^+ form). Evidently, the removal of contaminants during the first cationic exchange resin chromatography, as well as the acidic components adhering to the Dowex-1 resin, was sufficient to permit effective purification by repeating the Dowex-50 (H^+) step. *cis*-α-(Carboxycyclopropyl)glycine exchanged with both the cationic and anionic exchange resins; thus, it is not altogether surprising that these workers turned to a distinctive process such as partition chromatography to complete the purification. It would seem that the contaminant compounds possessed ion-exchange properties close to that of the desired compound.

The above procedures illustrate that successful purification of a particular nonprotein amino acid necessitates an appreciation of both the fundamental amphoteric properties of the sought after substance and the types and nature of the contaminant compounds from which it must be purified. Each particular situation must be individually tailored to achieve the desired aims. Generalizations beyond a certain point are not only meaningless but they hamper the capacity of the investigator to maximize the strong points of a particular system.

The need for flexibility in applying ion-exchange chromatographic procedures to individual situations is illustrated aptly in the imaginative procedure for the preparation from natural sources of L-hypoglycin free of leucine and isoleucine (Fowden, 1975). Standard procedures, even repeated crystallizations, failed to remove the contaminant protein amino acids. Fowden applied a hypoglycin-containing seed extract to a column containing Dowex-1 resin. Neutral amino acids, including the leucines, as well as hypoglycin elute from the first resin with the aqueous wash; the γ-glutamyl derivative of hypoglycin, also stored in the seed, as well as acidic seed components are retained by the resin.

The γ-glutamyl derivative is eluted with acetic acid, hydrolyzed with formic acid, and the hydrolysate (containing the newly produced hypoglycin) is rechromatographed as above. In the second chromatographic separation, unreacted formic acid and the acidic components contained in the original acetic acid-containing developing solvent are now retained on the resin. As before, hypoglycin is eluted with the aqueous wash. By these procedures, the bulk of the undesirable seed components are washed from the column with hypoglycin; the γ-glutamyl dipeptide and the acidic components are retained. These acidic components are subsequently removed once hypoglycin is generated from its γ-glutamyl derivative by acid hydrolysis. In this way, leucine-free hypoglycin is secured.

4. Automated Amino Acid Analysis

Automated amino acid analysis arose primarily through the pioneering efforts of S. Moore and W. H. Stein of the Rockefeller Institute of New York. It is in the area of analytical amino acid technology that automated ion-exchange chromatography has had its most dramatic impact. Amino acid compositional data are now a routine aspect of enzyme purification and characterization. This methodology has also found numerous biomedical and general research applications.

Essentially, automated amino acid analysis consists of an integrated system for pumping a series of appropriate buffers onto an ion-exchange resin-containing column. Bound amino acids are eluted sequentially from the column and mixed with independently pumped ninhydrin. The amino acid–ninhydrin mixture then traverses through a heating coil for color development. A dual wavelength recording spectrophotometer or two separate spectrophotometers designed to measure absorbance at 570 and 440 nm assay the resulting colors. The former wavelength responds to the Ruhemann's purple generated by most ninhydrin-treated amino acids while the latter effectively quantitates the yellow color of proline, hydroxyproline, and similar constituents. Output from the spectrophotometer need only be connected to a two-channel millivolt recorder to obtain simultaneous 570 and 440 nm tracings. Single wavelength detection of both the amino and imino acids of a sample is possible, e.g., at 420 nm (Rokushika *et al.*, 1977).

Automated amino acid analysis has also proven a valuable research tool for nonprotein amino acid studies although most of the analytical

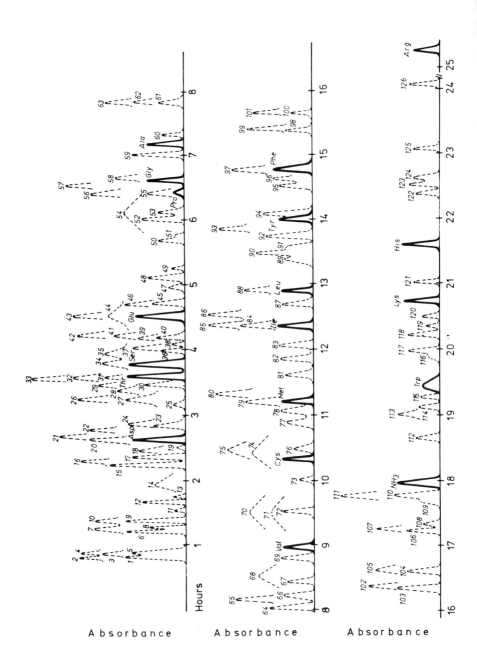

work was conducted when paper partition chromatography was in vogue. Single column, high pressure amino acid analyzers designed specifically for evaluating complex physiological fluids or nonprotein amino acid mixtures are available commercially. Automated amino acid analysis of complex mixtures, of course, requires a longer run time than does a protein hydrolysate. While the latter can be achieved routinely in

Fig. 7. Elution positions of 145 ninhydrin-positive compounds. The 18 physiological amino acids and ammonia are indicated by solid lines. The remaining compounds are 1, cysteic acid; 2, homocysteic acid; 3, cysteinesulfinic acid; 4, O-phosphothreonine; 5, O-phosphoserine; 6, taurine; 7, penicillaminic acid; 8, *threo*-β-hydroxyaspartic acid; 9, phosphoethanolamine; 10, levulinic acid; 11, dithiothreitol; 12, *erythro*-β-hydroxyaspartic acid; 13, urea; 14, S-carboxymethylglutathione; 15, allo-γ-hydroxyglutamic acid; 16, S-methylcysteine sulfoxide; 17, allo-β-hydroxyglutamic acid; 18, cephalosporin C; 19, 3-hydroxypipecolic acid; 20, S-carboxymethylcysteine; 21, S-carboxymethylpenicillamine; 22, diaminosuccinic acid (peak 1); 23, glutathione (reduced); 24, S-methylglutathione; 25, 4-hydroxyproline; 26, diaminosuccinic acid (peak 2); 27, penicilloic acid of isopenicillin N; 28, γ-(L-glutamyl)-L-cysteine; 29, methionine sulfoxide (peak 1); 30, methionine sulfone; 31, methionine sulfoxide (peak 2); 32, allothreonine; 33, β-hydroxyvaline; 34, δ-(L-α-aminoadipyl)-L-cysteine; 35, δ-(L-aminoadipyl)-L-cysteinyl-L-valine; 36, O-methylthreonine; 37, O-methylserine; 38, allo-4-hydroxyproline; 39, muramic acid; 40, asparagine; 41, allo-4-hydroxypipecolic acid; 42, S-carbamylcysteine; 43, β-methoxyvaline; 44, δ-(L-α-amino-adipyl)-L-cysteinyl-D-valine; 45, glutamine; 46, homoserine; 47, 4-oxopipecolic acid; 48, sarcosine; 49, 5-hydroxypipecolic acid; 50, cysteine; 51, *threo*-thiolbutyrine; 52, S-methylcysteine; 53, α-aminoadipic acid; 54, glutathione (oxidized); 55, *erythro*-thiolbutyrine; 56, S-carboxymethylhomocysteine; 57, β-hydroxyleucine; 58, penicillamine (reduced); 59, isoserine; 60, lanthionine (peak 1); 61, citrulline; 62, lanthionine (peak 2); 63, α-aminoisobutyric acid; 64, glucosamine; 65, S-ethylcysteine; 66, α-aminobutyric acid; 67, mannosamine; 68, bis-γ-(L-glutamyl)-L-cystine; 69, galactosamine; 70, bis-δ-(L-amino-adipyl)-L-cystine; 71, bis-δ-(L-α-aminoadipyl)-L-cystinyl-bis-L-valine; 72, α-aminopimelic acid; 73, pipecolic acid; 74, bis-δ-(L-α-aminoadipyl)-L-cystinyl-bis-D-valine; 75, 6-amino-penicillanic acid; 76, homocysteine; 77, phenylglycine; 78, homocitrulline; 79, norvaline; 80, mixed disulfide of L-cysteine and D-penicillamine; 81, alloisoleucine; 82, ethionine; 83, djenkolic acid; 84, penicillamine (oxidized); 85, cystathionine; 86, allocystathionine; 87, α-amino-β-hydroxybutyric acid; 88, 3,4-dihydroxyphenylalanine; 89, isoglutamine; 90, α,ϵ-diaminopimelic acid; 91, norleucine; 92, cycloserine; 93, α-amino-β-ethylvaleric acid; 94, mixed disulfide of L-cysteine and DL-homocysteine; 95, β-alanine; 96, mixed disulfide of DL-homocysteine and D-penicillamine; 97, O-benzylserine; 98, β-aminoiso-butyric acid; 99, δ-aminolevulinic acid; 100, L-cysteinyl-L-valine; 101, L-cysteinyl-D-valine; 102, argininosuccinic acid; 103, homocystine; 104, γ-aminobutyric acid; 105, S-benzyl-cysteine; 106, 5-hydroxytryptophan; 107, α-aminocaprylic acid; 108, ethanolamine; 109, kynurenine; 110, L-cystinylbis-L-valine; 111, L-cystinylbis-D-valine; 112, δ-aminovaleric acid; 113, valinol; 114, 5-hydroxylysine; 115, allo-5-hydroxylysine; 116, creatinine; 117, α,γ-diaminobutyric acid; 118, ornithine; 119, valinamide; 120, ϵ-aminocaproic acid; 121, 1-methylhistidine; 122, 3-methylhistidine; 123, carnosine; 124, homocarnosine; 125, α-amino-β-guanidinopropionic acid; 126, homocysteine thiolactone. Reprinted with permission from Adriaens *et al.* (1977). Copyright by the Elsevier Scientific Publishing Company. Reproduced from a drawing provided by B. Meesschaert.

90 min, analysis of 3 to 4 hr and occasionally longer are needed for effective resolution of nonprotein amino acid mixtures (Fig. 7).

Commercial integrators are available which automatically determine the area of the eluted peaks. Numerous statistical parameters can be applied automatically to the information processed by these calculating integrators and they are also capable of discounting such artifacts as base line irregularities.

5. Amino Acid Detection

In standard amino acid analysis, compounds are detected concurrently at 570 and 440 nm. There is, of course, no reason for limiting the analytical beam solely to these wavelengths since many nonprotein amino acids possess other absorption maxima for their ninhydrin–amino acid complex. Charlwood and Bell (1977) extended this concept and employed multiple spectrophotometers as a means of comparing the *ratio* of the absorbance of the Ruhemann's purple formed after 3 min at 570 nm to 405, 416, 434, and 475 nm as well as the A_{570} after 3 and 15 min of reaction time (Table 1). Output from these colorimeters has been computerized to create an informational bank in which the column retention time and absorbance ratios for each compound processed are recorded. Such catalogued data can be retrieved at will for comparison with more recently analyzed samples.

A major improvement in the sensitivity of amino acid detection has been achieved with the recent discovery that fluorescamine reacts at room temperature and alkaline pH with amino acids in several hundred milliseconds to produce a highly fluorescent product. Excess reagent does not interfere since it lacks appreciable fluorescence (Stein *et al.*, 1973). After the fluorescamine is dissolved in a suitable buffer, such as borate, it is mixed with the amino acid-containing effluent and passed to a spectrofluorometer. With this reagent, peak areas are created from information provided by a spectrofluorometer, which is able to detect the amino acid concentration with much greater sensitivity than is color-level determination with a spectrophotometer. Fluorescamine-based analysis of amino acids permits detection of as little as 50 to 100 pmoles of amino acid, a particular benefit when sample material is limited. In addition, the reactivity of the reagent negates the need for a heating coil and a boiling water bath. On the other hand, this reagent does not permit direct detection of imino acids since pyrrolinones, the highly fluorescent compounds produced from fluorescamine, form only with primary amines. This fluorimetric assay can be rendered sensitive to imino acids by prior reaction with N-chlorosuccinimide or hypochlo-

Table 1 Wavelengths of Absorption Maxima and Absorbance Ratios of Ninhydrin Reaction Products of Various Amino Acids[a]

Amino acid	Wavelengths of maxima (nm)	Absorbance ratio					
		$\dfrac{A_{405}\ (3\ \text{min})}{A_{570}\ (3\ \text{min})}$	$\dfrac{A_{416}\ (3\ \text{min})}{A_{570}\ (3\ \text{min})}$	$\dfrac{A_{434}\ (3\ \text{min})}{A_{570}\ (3\ \text{min})}$	$\dfrac{A_{475}\ (3\ \text{min})}{A_{570}\ (3\ \text{min})}$	$\dfrac{A_{570}\ (3\ \text{min})}{A_{570}\ (15\ \text{min})}$
Alanine	405, 570	1.05	0.79	0.21	0.13	0.11
Tryptophan	405, 570	1.00	0.76	0.50	0.35	0.45
Proline	415	7.21	7.12	6.48	4.54	0.69
4-Hydroxyproline	393, 447	8.43	7.86	7.07	4.71	0.65
Pipecolic acid	405, 432, 570	0.94	0.77	0.34	0.26	0.51
4-Hydroxypipecolic acid	438	6.08	12.17	17.00	11.83	0.58
4-Methylglutamic acid	405, 570	1.35	1.03	0.15	0.05	0.39
4-Methyleneglutamic acid	408, 470, 570	2.89	2.78	2.89	3.22	0.06
4-Ethylideneglutamic acid	405, 462, 570	1.10	0.90	0.31	0.31	0.51
Homoarginine	405, 570	1.05	0.75	0.12	0.06	0.56
3,4-Dihydroxyphenylalanine	405, 570	1.11	0.82	0.14	0.07	0.86
2,4-Diaminobutyric acid	405, 570	1.13	0.88	0.15	0.10	0.47
2-Amino-3-methylaminopropionic acid	405, 570	1.08	0.89	0.39	0.39	0.50
N-Methyltyrosine	405, 570	1.00	0.75	0.13	0.13	0.44
Lathyrine	415, 470	6.33	6.56	6.67	6.56	1.00
Canavanine	405, 570	1.18	0.85	0.18	0.09	0.85
Albizziine	405, 570	1.05	0.81	0.13	0.06	0.50
3-Carboxy-6,7-dihydroxy-1,2,3,4-tetrahydroisoquinoline	398, 570	2.20	1.60	0.50	0.01	0.30
Dichrostachinic acid	405, 570	1.11	0.79	0.14	0.08	0.57
Djenkolic acid	405, 456, 570	1.15	0.90	0.26	0.18	0.78
Azetidine-2-carboxylic acid	417, 570	1.67	1.33	0.52	0.38	0.78
Mimosine	405, 570	1.09	0.78	0.12	0.07	0.88

[a] The ratios of the absorbance of a ninhydrin reaction product at different wavelengths after a 3-min reaction time give an indication of the color of the product mixture; the ratio of absorbances at 570 nm after 3- and 15-min development is a measure of the rate of production of the color. Reproduced with permission from the work of Charlwood and Bell (1977). Copyright by Elsevier Scientific Publishing Company.

rite. However, the need for an oxidizing agent to produce fluorogenic reaction products increases the complexity of the detection system (Roth and Hampaï, 1973; Weigele *et al.*, 1973). Fluorescamine in a suitable organic vehicle such as acetone or dimethylsulfoxide can be applied to *in situ* detection of amino acids separated by thin-layer chromatography (Touchstone *et al.*, 1976).

Fluorescent detection of amino acids in the picomole range is obtainable with other reagents, for example, *O*-phthalaldehyde (Benson and Hare, 1975). This fluorescent reagent in conjunction with 2-mercaptoethanol can be used for the detection of amino acids, amines, as well as peptides, and has been commonly applied to compound detection in thin-layer chromatography (Lindeberg, 1976). Amino acids react with pyridoxal under alkaline conditions to form a Schiff base that can be reduced chemically with tetrahydroborate to produce pyridoxyl–amino acid derivatives. These derivatives can be detected at a concentration of $5 \times 10^{-10} M$ and exhibit an excitation maxima (λ_{ex}) of 322 nm and an emission maxima (λ_{em}) of 440 nm (Lustenberger *et al.*, 1972; Lange *et al.*, 1972). In a modification of this procedure, pyridoxal phosphate and zinc acetate, dissolved in pyridinemethanol, interact with amino acids to form easily measured fluorophors. These fluorogenic derivatives have a λ_{ex} of 390 ± 5 nm and a λ_{em} of 470 ± 5 nm. Neither tryptophan nor proline produce a fluorescent derivative by this procedure (Maeda and Tsuji, 1973).

Certain sulfonyl chlorides react with amino acids to form fluorogenic derivatives. Among the best known of this group are the fluorescent derivatives of 5-dimethylaminonaphthalene-1-sulfonyl chloride (commonly known as dansyl chloride). Sulfonyl chlorides are noteworthy for yielding detectable fluorescence with nanogram amounts of amino acids (see Gray and Smith, 1970; Seiler, 1970; and Deyl and Rosmus, 1972). When dansylated amino acids are separated by thin-layer chromatography and eluted with a suitable solvent, they can be analyzed with a spectrofluorometer. Airhart *et al.* (1973) claim a 1000-fold enhancement in detection capability for the dansyl derivative procedure as compared to routine amino acid analysis.

Cyanate is a well-established carbamylating agent of the α-NH_2 group of amino acids. This reaction has also been employed widely in the identification of the N-terminal residue of proteins by sequential carbamylation and acidification to produce the corresponding hydantoin (see Stark, 1967). Amino acid hydantoin derivatives, separated by thin-layer chromatography, can be detected with *tert*-hypochlorite (Suzuki *et al.*, 1973). Use of various sulfonyl chlorides or the production of the hydantoin or pyridoxal derivatives has not been applied extensively to nonprotein amino acid detection; they may represent a type of analytical methodology of value.

F. GAS CHROMATOGRAPHY

Gas chromatography or more correctly gas–liquid chromatography is a separational process based on compound partitioning in an admixture between the gaseous carrier phase and a supported liquid phase. An inert gas, e.g., nitrogen, helium, or argon, at a pressure of 2 to 3 atm is used to flush an injection device containing the sample which is either preheated or simply delivered to the chromatographic column.

The chromatographic column is made of various materials including glass, stainless steel, or a polymeric substance such as polythene. Analytical columns are generally 1 to 3 m long with an internal diameter of 2 to 6 mm; longer columns are coiled for more effective heating. Preparative columns have a much larger diameter (3 cm or more) and these can extend over 100 m in length. Throughout the chromatographic run, the column is maintained at a constant temperature, generally ranging from 50° to 500°C, but provisions usually exist for programming changes in the column operating temperature.

Within the column, the various components of the sample are partitioned continuously between the gaseous carrier phase and the supported liquid phase. The differentially eluted substances are then transported to a detector which generates a signal in proportion to the concentration of the eluted substance. After amplification, these signals are expressed visually as distinctive "peaks." The volume of the carrier gas which passes from the point of sample injection to the visualization of the peak maximum is known as the retention volume and is wholly analogous to the retention time of ion-exchange chromatography.

1.   Sample Preparation

Amino acids being polar, amphoteric substances are characterized by strong electrostatic interaction which not only imparts a high melting point but also a low vapor pressure. For this reason, the volatility of these compounds must be amplified prior to chromatographic separation. Any method for converting amino acids to volatile derivatives must involve quantitative formation of the derivatives, should be applicable to amino acids of differing structure, and must not create volatile contaminants. Many reactions achieve these basic aims and share in common the eventual formation of a more volatile substance, usually by effectively blocking the amino acid's polar groups. In an early approach to resolving this need, the carboxyl group was simply esterified with acidified short chained aliphatic alcohols; these simple esters have given way to silylated derivatives. In the process taken to represent silylation of the amino acid, the proton is replaced with a trimethylsilyl group. The carboxyl group silylates readily and can be reacted with mild silylating agents such as hexamethyldisilazane or trimethylchlorosilane. Amino group masking can also be achieved but stronger silylating conditions are required.

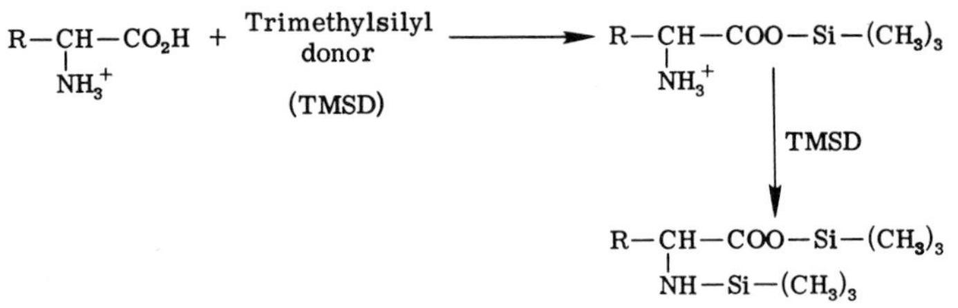

$$\underset{\overset{|}{NH_3^+}}{R-CH-CO_2H} \; + \; \overset{\text{Trimethylsilyl}}{\underset{\text{(TMSD)}}{\text{donor}}} \longrightarrow \underset{\overset{|}{NH_3^+}}{R-CH-COO-Si-(CH_3)_3}$$

$$\Big\downarrow \text{TMSD}$$

$$\underset{\overset{|}{NH-Si-(CH_3)_3}}{R-CH-COO-Si-(CH_3)_3}$$

Many silylating agents are more reactive than trimethylsilylic acid; for example, trimethylsilyldiethylamine facilely silylates both polar groups simultaneously, and this is presently one of the most effective silylating agents.

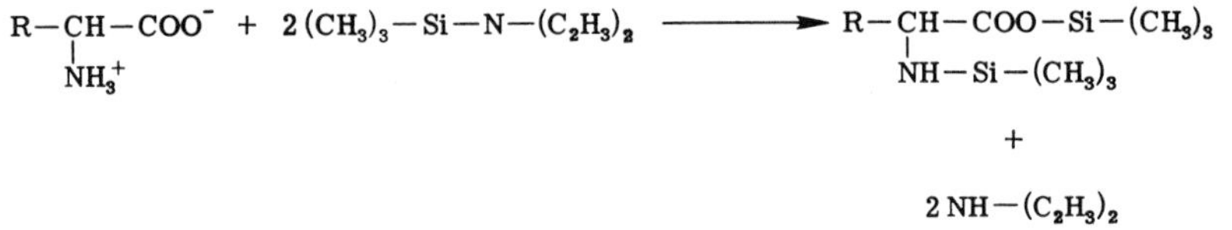

$$\underset{\overset{|}{NH_3^+}}{R-CH-COO^-} \; + \; 2\,(CH_3)_3-Si-N-(C_2H_5)_2 \longrightarrow \underset{\overset{|}{NH-Si-(CH_3)_3}}{R-CH-COO-Si-(CH_3)_3}$$

$$+$$

$$2\,NH-(C_2H_5)_2$$

Many amino acid derivatives have yielded to acceptable chromatographic separation and these can be organized in terms of their masked α-amino and α-carboxyl group. Thus, beside the fully silylated derivatives, variously combined esters such as N-trimethylsilyl amino acid ethyl esters are also suitable derivatives. Acylation has also proved a useful blocking procedure. Generally, the amino acids are esterified initially with low molecular weight aliphatic alcohols and then trifluoroacetylated with trifluoroacetic anhydride in methylene chloride. In this way, highly volatile compounds such as *n*-butyl-N-trifluoroacetyl derivatives are easily prepared. The methyl esters of the N-trifluoroacetyl amino acids were among the most highly regarded for amino acid analysis, but they are somewhat ephemeral and recently have given way to the 1-pentyl or 1-butyl derivatives. The 1-butyl esters of 24 nonprotein amino acids have been chromatographed in the presence of 14 protein amino acids with excellent compound separation in under 1 hr (Fig. 8) (see also Raulin et al., 1972). Simple acetylation by treatment with acetic anhydride in pyridine also represented an effective group masking strategy. In fact, selenium-containing amino acids chromatograph particularly well as their n-propyl-N-acetyl derivatives (Coulter and Hann, 1971).

Recently, a group of diverse nonprotein amino acids were reacted with isobutyl chloroformate in aqueous solution followed by esterification with diazomethane. These procedures yielded a group of novel compounds, namely N-isobutyloxycarbonyl methyl esters, which can be resolved in less than 1 hr (Fig. 9) with excellent reproducibility between runs (Makita et al., 1976).

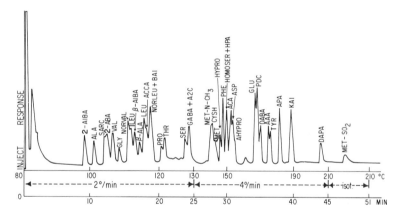

Fig. 8. Gas–liquid chromatography of 24 nonprotein and 14 protein amino acids. The initial temperature was 60°C with a programmed rate increase of 2°C min⁻¹ for 25 min and then 4°C min⁻¹ for 20 min until a final temperature of 210°C was attained. 2-AIBA, α-Aminoisobutyric acid; ALA, alanine; SARC, sarcosine; 2-ABA α-aminobutyric acid; VAL, valine; GLY, glycine; NORVAL, norvaline; ILEU, isoleucine; β-AIBA, β-aminoisobutyric acid; β-ALA, β-alanine; LEU, leucine; ACCA, 1-amino-1-cyclopropanecarboxylic acid; NORLEU, norleucine; BAI, baikiain; PRO, proline; THR, threonine; SER, serine; GABA, γ-aminobutyric acid; A2C, azetidine-2-carboxylic acid; MET-N-CH₃, N-methyl-methionine; CYSH, cysteine; MET, methionine; HYPRO, hydroxyproline; PHE, phenyl-alanine; HOMOSER, homoserine; HPA, 5-hydroxypipecolic acid; ACA, ε-aminocaproic acid; ASP, aspartic acid; AHYPRO, allohydroxyproline; GLU, glutamic acid; PDC, pyrrolidine-2,5-dicarboxylic acid; DABA, 2,4-diaminobutyric acid; AAA, α-aminoadipic acid; TYR, tyrosine; APA, α-aminopimelic acid; KAI, kainic acid; DAPA, α,α-diamino-pimelic acid; MET-SO₂, methionine sulfone. Reprinted with permission from Amico et al. (1976). Copyright by the Elsevier Scientific Publishing Company.

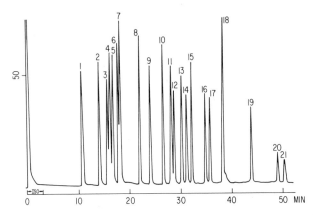

Fig. 9. Chromatographic separation of the *n*-isobutyloxycarbonyl methyl esters of certain nonprotein amino acids. Each peak represents 2 μg of amino acid. 1, Sarcosine; 2, α-amino-*n*-butyric acid; 3, alloisoleucine; 4, norvaline; 5, β-aminoisobutyric acid; 6, β-alanine; 7, norleucine; 8, γ-aminobutyric acid; 9, *S*-methylcysteine; 10, ε-aminocaproic acid; 11, ethionine; 12, homoserine; 13, α-aminoadipic acid; 14, δ-aminolevulinic acid; 15, kainic acid (internal standard); 16, *S*-carboxymethylcysteine; 17, homocysteine; 18, 2,4-diaminobutyric acid; 19, methionine sulfone; 20, lanthionine; 21, δ-hydroxylysine. Reprinted with permission from Makita *et al.* (1976). Copyright by the Elsevier Scientific Company.

2. Column Supports and Liquid Phases

Diatomaceous earths are among the most widely used solid media for packing columns. These materials, however, must first be washed sufficiently with mineral acids to leach metallic ions and then treated chemically with silanizing agents such as dimethylchlorosilane to neutralize reactive surface groups by blocking their adsorptive capacity. This encourages unimpeded partition of carrier gas components into the liquid phase rather than their being held adsorptively by the support medium. If the chromatographic process proceeds adsorptively in addition to the usual partitional modes, an asymmetric elution pattern inevitably results.

Once diatomaceous earths are treated properly, their ability to resolve various compounds, such as the *n*-propyl-*N*-acetyl amino acids, is vastly improved. They possess a large reactive surface with a strong affinity for the liquid phase, resist loss during packing, and can be rendered inert. Chromosorb P and W are highly regarded commercial preparations of diatomaceous earths; glass beads and various Teflon powders constitute other suitable solid supports. The separational efficacy of these solid supports is augmented significantly by decreased particle size since this enlarges the reactive surface area. Beyond a certain point, unacceptable pressure loads and protracted run times become undesirable consequences of continued particle size reduction.

A large number of liquid components have functioned as the supported stationary phase in gas–liquid chromatography. Among the better regarded are the silicones, polyesters, and polyglycols. The selected substances must be stable, inert, relatively nonvolatile, and they must not contribute matter to the carrier gas. Apolar materials generally elute from an apolar stationary phase in relation to their respective boiling point. As the polarity of the stationary phase is extended, the retention volume for polar substances is also increased. Generalizations can only be of limited value; once again a trial and error approach is unavoidable. The interested reader is directed to the informational sources provided at the end of this chapter for data on various solid and liquid phases and their operational characteristics.

3. Detectors

Once the derivatized amino acids elute from the column, a means must be available for their detection. The detection process is predicated fundamentally on the alteration, by components of the column effluent, on some measurable parameter of the carrier gas, for example, the thermal conductivity, heat of combustion, or density of the carrier gas. Factors not related to the presence of column components can affect the nature of the carrier gas; when this occurs it creates fluctuations in the base line which appear as anomalous "peaks." As a result, the detector system must not only be sensitive and accurate but must minimize the formation of such spurious peaks. The detector must respond rapidly, in a linear manner, and be able to resist perturbation caused by changes in pressure, temperature, or carrier gas flow rate.

The flame ionization apparatus is illustrative of present day detectors. A hydrogen flame burning in air or oxygen possesses an electrical conductivity of only 10^{-12} amperes. When the column effluent enters the flame, the chemical vapors dramatically increase the electrical conductivity. The resulting sharp rise in conductivity is proportional to the rate of solute ingress into the flame detector (Morris and Morris, 1973). In a simple modification of this procedure, a sensitive thermocouple is employed to evaluate the temperature change that occurs in a flame of burning hydrogen. When a foreign substance enters the flame, its heat of combustion is detected and quantified.

Many other types of detectors have been employed successfully but one has particular value in amino acid analysis, i.e., a detector based on electron capture techniques. Since halogens efficaciously trap electrons, halogenated amino acid derivatives, such as trifluoroacetyl compounds, are particularly amenable to this type of detection system.

In conclusion, Coulter and Hann (1971) list four chief advantages of gas-liquid over ion-exchange chromatography. These include greater sensitivity, reduced analysis time, decreased cost, and greater utility. The final factor reflects its greater potential for application in the analysis of natural products other than amino acids. Certain disadvantages also exist, for example, greater sample preparation time since the amino acids must be quantitatively derivatized and greater difficulty in maintaining constant and reproducible operating conditions. Nevertheless, gas-liquid chromatography is gaining an increasing number of converts and it (as well as liquid-liquid chromatography) may some day be of far greater use in the analysis of amino acids.

G. ELECTROPHORETIC TECHNIQUES

The zwitterion nature of amino acids creates a mechanism for effective separation of these natural products based on their differential migration in an electrical field. In this process, the sample mixture in a suitable buffer is applied as a narrow circular zone or broad band at the center of the selected supporting matrix. Electrophoresis, i.e., the migration of materials within an electric field, is conducted with a suitable solvent system that functions as the liquid phase. Paper, starch, cellulose powder or acetate, and silica gel commonly constitute the supporting medium, and are saturated with the liquid phase prior to applying a constant direct current across the supporting medium. Application of this uniform electrical field causes the charged amino acids to move. They rapidly attain a constant migrational rate that is a function of many parameters including the temperature, magnitude of the generated electrical field, as well as the ionic strength and viscosity of the solvent system. Molecular size and shape also affect molecular travel rates since these properties alter the frictional and electrostatic forces exerted by the surrounding medium.

With amino acids, the pH of the liquid phase has a preponderant influence on compound electrophoretic displacement. Consider glycine with its pK_a values of 2.35 and 9.78; at its isoelectric point of 6.06 this amphoteric molecule exists overwhelmingly as the neutral species and movement results essentially from electroosmotic flow through the medium. At pH conditions above 6.06, it is more anionic than at the isoelectric point; below 6.06 it is more cationic.

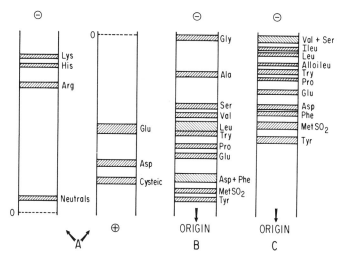

Fig. 10. High-voltage electrophoretic separation of various amino acids. (A) Pyridine-acetate buffer, pH 5.2; (B) formic acid–acetic acid; pH 1.85; (C) formic acid–acetic acid–calcium acetate, pH 1.81. Reprinted with permission from Atfield and Morris (1961). Copyright by the Biochemical Society.

As shown in Fig. 10, at a pH of 5.2, acidic, basic, and neutral members segregate well from one another but resolution within neutral members is poor. On the other hand, at a pH below 2 where amino acids are highly protonated, migration is increased greatly and this facilitates separation of previously unresolved amino acids (Fig. 10). The separation depicted in Fig. 10 can be achieved by the use of an electrophoretic solvent consisting of 20 gm pyridine and 9.5 gm of acetic acid/liter (pH 5.2) for a 200 min run at 75 V cm^{-1}. This procedure resolves the basic amino acids as a cationic group while the acidic components move as anions. The neutral members remain largely unresolved as a group which has migrated 1.5 to 2 cm toward the cathode by osmotic flow over the paper (Zweig and Whitaker, 1967). Basic natural products electrophorese more effectively at higher pH—generally around 5.5 to 7.5; but the use of alkaline pH systems has not gained widespread acceptance. An effective regime for the separation of basic components is provided by pyridine–acetic acid (pH 6.5) followed by 50 mM pH 11.2 borate buffer.

It is common practice to resolve amino acid mixtures by electrophoresis in one direction using a volatile solvent system such as formic acid–acetic acid–water or pyridine–acetic acid and a field strength of at least 50 V cm^{-1}. After drying the chromatogram, it is subjected to partition chromatography at an angle of 90° to the first direction. The resolution achieved by combined electrophoresis and partition chromatography can be quite striking with separation of certain diastereoisomers being easily resolved. Electrophoresis can, of course, be conducted in both directions employing chemically distinctive liquid phases. The use of higher voltages (up to 100 V cm^{-1}) reduces significantly the run time which in turn minimizes diffusion and thereby enhances overall resolution; increasingly, such high operating voltages are becoming routine.

As a beginning, two-dimensional high voltage electrophoresis of a nonprotein amino acid mixture can be conducted with pH 2.0 formic acid in the first direction after introducing the sample near the edge of the supporting medium adjacent to the anode. Electrophoresis at 100 V cm^{-1} for 40 min is suggested for the initial separation. Sodium borate buffer (50 mM at pH 9.2) is recommended for a second run of 20 min duration at the same operating potential. Cellulose acetate or silica gel should be considered as an alternative to paper; media of many diverse compositions are available commercially.

Isolation and the subsequent purification of nonprotein amino acids by electrophoresis is limited customarily to several milligrams of sample but it is possible to greatly increase the yield. In continuous electrophoresis, electrophoretic separation is conducted at a right angle to the sustained flow of the solvent system across a stationary phase such as paper (Fig. 11). The edge of the paper distal to sample application is serrated and the material cascading from each serrated segment is collected as an individual fraction. In this way, gram quantities of material can be processed routinely.

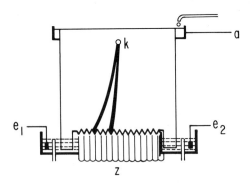

Fig. 11. Durrum apparatus for continuous electrophoresis. a, Buffer-containing trough; k, sample feed, e_1 and e_2, electrodes; z, receivers for collecting fractions. Reprinted with permission from Mikeš (1966). Copyright by D. Van Nostrand Company, Ltd.

H. OTHER ANALYTICAL TECHNIQUES

Several other analytical techniques are currently enjoying intense interest and development. High performance liquid chromatography is viewed by many as the analytical technique of the future particularly as new and useful derivatization procedures are developed; its application to nonprotein amino acid analysis is just beginning. A new approach to amino acid identification and characterization involves the breakup of the molecule into fragments which can be characterized according to their mass. The use of *mass fragmentography* permits the identification of 10^{-10} gm amounts of a compound (Iwase and Murai, 1977). Since the amino acid is derivatized prior to fragmentation, samples previously separated by gas chromatography can be subjected directly to mass analysis as they emerge with the carrier gas stream. This combined use of gas chromatography and mass spectroscopy is finding ever-increasing application in studies involving heavy isotope incorporation into amino acids.

ADDITIONAL READING

Bier, M., ed. (1967). "Electrophoresis. Theory, Methods, and Application," Vol. 2. Academic Press, New York.

Blackburn, S. (1965). The determination of amino acids by high voltage paper electrophoresis. *Methods Biochem. Anal.* **13**, 1–45.

Blackburn, S. (1968). "Amino Acid Determinations." Dekker, New York.

Block, R. J., Durrum, E. L., and Zweig, G., eds. (1958). "A Manual of Paper Chromatography and Paper Electrophoresis," 2nd ed. Academic Press, New York.

Bobbitt, J. M. (1963). "Thin-Layer Chromatography." Van Nostrand-Reinhold, Princeton, New Jersey.

Brenner, N., Callen, J. E., and Weiss, M. D., eds. (1962). "Gas Chromatography." Academic Press, New York.

Burchfield, H. P., and Storrs, E. E. (1962). "Biochemical Applications of Gas Chromatography." Academic Press, New York.

Devenyi, T., and Gergely, J. (1974). "Amino Acids, Peptides and Proteins." Am. Elsevier, New York.

Done, J. N., and Knox, J. H. (1974). "Applications of High-Speed Liquid Chromatography." Wiley, New York.

Florkin, M., and Stotz, E. H., eds. (1962). "Comprehensive Biochemistry," Vol. 4. Am. Elsevier, New York.

Fowler, L., ed. (1963). "Gas Chromatography." Academic Press, New York.

Hais, I. M., and Macek, K., eds. (1963). "Paper Chromatography: A Comprehensive Treatise." Academic Press, New York.

Heftmann, E., ed. (1967). "Chromatography," 2nd ed. Van Nostrand-Reinhold, Princeton, New Jersey.

Husek, P., and Macek, K. (1975). Gas chromatography of amino acids. *J. Chromatogr.* **113**, 139–230.

Liteanu, C., and Gocan, S. (1974). "Gradient Liquid Chromatography." Wiley, New York.

Littlewood, A. B. (1962). "Gas Chromatography. Principles, Techniques, and Applications." Academic Press, New York.

Mikeš, O., ed. (1961). "Laboratory Handbook of Chromatographic Methods." Van Nostrand-Reinhold, Princeton, New Jersey.

Mitruka, B. M. (1975). "Gas Chromatographic Applications in Microbiology and Medicine." Wiley, New York.

Niederwieser, A., and Pataki, G., eds. (1971). "New Techniques in Amino Acid, Peptide, and Protein Analysis." Ann Arbor Sci. Publ., Ann Arbor, Michigan.

Pataki, G., ed. (1968). "Techniques of Thin-Layer Chromatography in Amino Acids and Peptide Chemistry." Ann Arbor Sci. Publ., Ann Arbor, Michigan.

Pataki, G., ed. (1970). "Techniques of Thin-Layer Chromatography in Amino Acid and Peptide Chemistry," 2nd ed. Ann Arbor-Humphrey Sci. Publ., Ann Arbor, Michigan.

Porter, R., ed. (1969). "Gas Chromatography in Biology and Medicine." J. & A. Church Ltd., London.

Purnell, H. (1962). "Gas Chromatography." Wiley, New York.

Sherma, J., and Zweig, G. (1971). "Paper Chromatography and Electrophoresis," Vol. 2. Academic Press, New York.

Smith, I., ed. (1969). "Chromatographic and Electrophoretic Techniques," Vols. 1 and 2. Wiley, New York.

Snyder, L. R., and Kirkland, J. J. (1974). "Introduction to Modern Liquid Chromatography." Wiley, New York.

Stahl, E., ed. (1969). "Thin-Layer Chromatography. A Laboratory Handbook," 2nd ed. Springer-Verlag, Berlin and New York.

Szymanski, H. A., ed. (1968). "Biomedical Applications of Gas Chromatography," Vol. 2. Plenum, New York.

Touchstone, J. C., ed. (1973). "Quantitative Thin Layer Chromatography." Wiley, New York.

Touchstone,, J. C., and Dobbins, M. E. (1978). "Practice of Thin-Layer Chromatography." Wiley, New York.

Walker, J. Q., Jackson, M. T., Jr., and Maynard, J. B. (1977). "Chromatographic Systems Maintenance and Troubleshooting." Academic Press, New York.

Zweig, G., and Whitaker, J. R. (1967). "Paper Chromatography and Electrophoresis," Vol. 1. Academic Press, New York.

Chapter 3

Toxic Constituents and Their Related Metabolites

Plants are not just food for animals, and animals are not just decorations on the vegetation. The world is not green. It is colored lectin, tannin, cyanide, caffeine, aflatoxin, and canavanine. And there is a lot of cellulose thrown in to make the mix even more inedible. Animals are not ambulatory bomb calorimeters. They starve, they ache, they abort, they vomit, they remember, they die, and they evolve.

Daniel H. Janzen, 1977

A. LATHYROGENS AND NEUROTOXINS

Lathyritic seeds have been an article of human diet for several millenia and reference to an ancient malady associated with their intake is found in the writings of Hippocrates, Pliny, and Galen. This affliction has come to be known as neurolathyrism, after the suggestion of Selye (1957). Neurolathyrism is induced by such higher plants as *Lathyrus sativus, L. sylvestris, L. cicera,* and *L. clymenum* and it is characterized by a general muscular weakness which in extreme cases produces irreversible paralysis. Death can be the ultimate manifestation of the ingestion of these seeds. While the lathyritic syndrome is exhibited by humans as well as our domesticated animals, higher animal susceptibility varies greatly. Horses and cattle are particularly sensitive; dogs, pigs, and rabbits less so; while rats are quite resistant.

Neurolathyrism is distinctive from a second pathological form termed osteolathyrism, for the latter involves aberrations of bone and mesenchymal tissues (Selye, 1957). The osteolathyritic condition can be induced readily in laboratory animals maintained on a diet of *L. odoratus* seeds but a variety of man-made compounds duplicate these skeletal anomalies (Fig. 12A and B). In addition to these two distinctive afflictions, the term angiolathyrism has been created recently to denote disorders of the vascular system that occur concomitantly with osteolathyrism. Barrow *et al.* (1974) have documented the inability to differentiate clinically the bone deformities of osteolathyrism from the vascular ramifications of angiolathyrism. Such *Lathyrus* species as *L. hirsutus* and *L. pusillus* can cause both neurolathyritic and osteolathyritic symptoms (Selye, 1957); this contrasts markedly with *L. odoratus* which is exclusively osteolathyritic in its mode of action.

In 1954, Schilling and Strong announced the isolation of a crystalline substance from *L. odoratus* that produces the characteristic skeletal anomalies of osteolathyrism and showed it to be γ-L-glutamyl-β-aminopropionitrile (see McKay *et al.*, 1954). The γ-glutamyl moiety does not contribute to the molecule's biological properties and it is now accepted that β-aminopropionitrile* is the active osteolathyritic factor. This toxic substance is present in those *Lathyrus* species able to induce osteolathyrism but absent from organisms that are either innocuous or associated exclusively with neurolathyrism.

The staggering dimension of neurolathyrism is illustrated aptly by data from a 1958 survey which revealed as many as 25,000 cases of neurolathyrism from a single district in India having a population of 634,000. To make matters worse, *L. sativus* can withstand adverse climatic conditions, which tends to increase this plant's prevalence during conditions of famine when individual susceptibility to these toxins intensifies (Sarma and Padmanaban, 1969).

Several compounds appear to be responsible for neurolathyrism: β-cyanoalanine and its γ-glutamyl dipeptide, α,γ-diaminobutyric acid and its N^γ-oxalyl derivative, and the N^β-oxalyl derivative of α,β-diaminopropionic acid. Several causes for the toxicity exhibited by these amino acids are considered in subsequent sections dealing specifically with these natural products. In addition, some insight has been gained into the biological action of β-aminopropionitrile which is responsible for osteolathyrism. Lysyl oxidase deaminates oxidatively the ϵ-amino group of the hydroxylysine and lysine residues of collagen as well as the lysine of elastin to their corresponding aldehyde. The δ-semialdehyde of lysine is known trivially as allysine. Reaction of the aldehyde group of

---

*The so-called "lathyrus factor" of the literature refers to the γ-L-glutamyl derivative of β-aminopropionitrile.

A

B

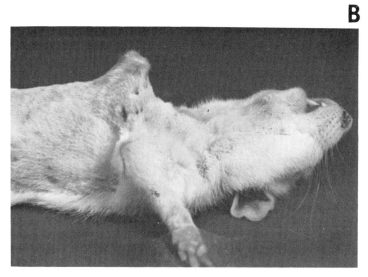

Fig. 12A. Skeletal deformity and aberrant cartilage formation in the rabbit. A young rabbit received 200 mg aminoacetonitrile/kg body weight per day for 10 days. This experimental compound mimics the action of β-aminopropionitrile. The skeletal deformity is

allysine with the ϵ-amino function of nonreacted lysine or hydroxylysine creates Schiff bases that form the covalent cross-linkages stabilizing collagen and elastin (O'Dell *et al.*, 1966; Piez, 1968; Siegel and Martin, 1970). The collagen and elastin found in β-aminopropionitrile-treated animals have a low allysine content (Siegel *et al.*, 1970) since β-aminopropionitrile effectively inhibits lysyl oxidase activity (Narayanan *et al.*, 1972). Such curtailment in allysine production prevents adequate formation of collagen–elastin cross-linkages which, in turn, yield connective tissues characterized by decreased tensile strength and excessively soluble collagen (Ressler, 1975).

To fully appreciate the potential biological impact of this toxic nitrile, remember that collagen is the most ubiquitous animal protein. It is of supreme importance in the formation and function of tendons and ligaments, skin and fascia, filtration structures as the glomeruli, supportive tissues such as bone and dentin, cartilage plus intervertebral discs, and specialized tissues such as those imparting light transmissional qualities to the cornea or fatigue resistance to the valves of the heart (Eyre, 1980).

1. β-Cyano-L-alanine

a. Background and Preparation

β-Cyanoalanine is distinctive in being the sole free non-protein amino acid of eukaryotes possessing the cyano group.

$$N{\equiv}C—CH_2—CH(NH_2)COOH$$

β-Cyano-L-alanine

The first preparation of this toxic nitrile was actually indirect, for it is a reaction intermediate in the chemical conversion of asparagine to α,γ-diaminobutyric acid (Ressler, 1975). Dehydration of the β-carboximide of asparagine produced β-cyanoalanine which was then reduced to create the desired diamino acid (Ressler and Ratzkin, 1961). These workers appreciated fully that decarboxylation of β-cyanoalanine forms β-aminopropionitrile, and considerable interest existed in the latter compound due to its role in osteolathyrism.

When this novel nitrile was evaluated biologically in the male rat by

---

sufficiently severe so that the animal is incapable of standing. A "softening" of the supporting cartilage is believed responsible for the lack of erectness of the ears.

Fig. 12B. Deformed xiphoid bone. The treated rat received 100 mg aminoacetonitrile/ kg body weight per day during a 6-week period. Reprinted with permission from Selye (1957). Copyright by Organe Officiel des Societes de Biologie. Photo provided by H. Selye.

administration of a 1% laboratory diet, it was found not to be associated with an expected osteolathyritic syndrome but rather to produce hyperirritability tremors, convulsions, and even death within 3 to 5 days (Ressler *et al.*, 1961). Discovery of its neurotoxic properties, in conjunction with the reports of Lewis *et al.* (1948) and Lewis and Schulert (1949) that *Lathyrus latifolius* and *L. sylvestris* are neurotoxic to rats and mice, led Ressler and associates to examine these plants for β-cyanoalanine. Their purification of the neurologically active fractions culminated in the unanticipated isolation of α,γ-diaminobutyric acid.

Ressler (1962) postulated subsequently that in *L. latifolius* and *L. sylvestris Wagneri* asparagine could be dehydrated to β-cyanoalanine prior to its reduction to α,γ-diaminobutyric acid. In *L. odoratus*, however, the amino acid nitrile is decarboxylated to β-aminopropionitrile (Fig. 13). It

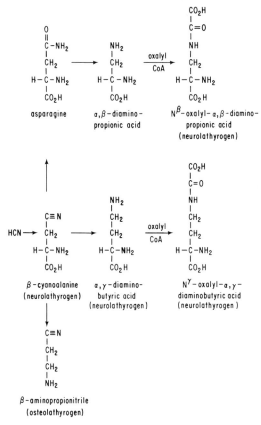

Fig. 13. Structural features and putative metabolic relationships of some secondary plant metabolites. From Rosenthal and Bell (1979).

was realized that human consumption of *L. sativus* was associated with neurolathyrism; yet, ingestion of the seeds failed to elicit any adverse biological effects. It was also known that *Vicia sativa* occurred frequently as a contaminating component of harvested *L. sativus;* consumption of the former seed generated a symptomology suggestive of a nervous disorder. Thus, she concentrated her search for the toxic factor in *V. sativa* and eventually isolated β-cyanoalanine. A survey of 48 species of *Vicia* disclosed this amino acid nitrile in 16 species (Bell and Tirimanna, 1965). Subsequently, it was obtained from *Chromobacterium violaceum* (Brysk *et al.*, 1969), which with selected *Vicia* and *Lathyrus* represent the currently known biological sources.

b. Biological Properties

Stomach tube administration of β-cyanoalanine (15 mg/100 gm fresh weight) to weanling male rats produced irreversible hyperactivity, tremors, convulsions, and rigidity while subcutaneous delivery of a slightly higher dose (20 mg/100 gm) evoked convulsions, rigidity, prostration, and death (Ressler, 1962). The level of β-cyanoalanine (0.15%) in *V. sativa* seemed to Ressler and associates insufficient to account for its observed toxicity in rat. Their astute observation instigated further effort that was rewarded ultimately with the successful isolation of an acidic dipeptide: γ-L-glutamyl-β-cyano-L-alanine. Its structure was confirmed both by chemical degradation of a derivative of the peptide that yielded a 1:1 ratio of glutamic acid to 2,4-diaminobutyric acid and by direct chemical synthesis. Some 0.6% of the seed dry weight but 1.7 to 2.6% of the young seedling (dry weight) is contributed by this toxic metabolite.

Subcutaneous application in rat suggested that on a molar basis, the dipeptide is as potent as the free amino acid; however, it is about one-half as toxic to White Leghorn chicks. The mode of administration is critical, since in a later report Ressler *et al.* (1969) stated: "In *feeding* experiments with several day-old chicks γ-glutamyl-β-cyanoalanine and free β-cyanoalanine have similar toxicities on a molar basis."

A mixture of the dipeptide and free amino acid, at one-half their level in the seed of *V. sativa* (0.075%), was incorporated subsequently into the young chicks' basal ration. One week later, these treated chicks entered into a terminal convulsive state and exhibited opisthotonus, i.e., tetanic spasms of the back muscles which force the head and lower limbs backward while the trunk arches forward (Fig. 14) (Ressler *et al.*, 1963). While nine times more amino acid nitrile is required to elicit neurotoxicity in the rat as compared to the chick, the LD_{50} in rat of 13.5 mg/100 gm is only about twice that for the avian (Ressler, 1975). Overall, Ressler *et al.* (1963) provided convincing evidence that β-cyanoalanine in conjunction

Fig. 14. β-Cyanoalanine-induced opisthotonus. The chick is experiencing a convulsive state with opisthotonus induced by this toxic nitrile. Reprinted with permission from Ressler *et al.* (1967). Copyright by Pergamon Press, Ltd. Photo provided by C. Ressler.

with its dipeptide derivative account for the neurotoxic properties of *V. sativa.*

This nitrile prevents significant growth of third stadium larvae of the locust *Locusta migratoria migratorioides* which perished subsequently without undergoing larval–larval ecdysis (Schlesinger *et al.*, 1976). Locust hemolymph volume decreases by 25% 1 day after a single hemolymphic injection (2 mg/insect). The volume of the hemolymph continues to decline precipitously until the fifth day when the larvae die. While this secondary metabolite undoubtedly alters many aspects of the insect's metabolism and physiology, these effects are overshadowed by the massive disruption of water balance. The ability of nonprotein amino acids to potentiate severe diuretic effects has been observed with other nonprotein amino acids, e.g., canavanine (see Rafaeli and Applebaum, 1980).

c. Toxicity

In their study of rat-β-cyanoalanine interaction, Ressler *et al.* (1964) noted that pyridoxal hydrochloride delayed initiation and reduced the severity of this amino acid's toxicity. They provided (subcutaneously) 20 to 23 mg β-cyanoalanine/100 gm body weight to weanling Sherman rats and a comparable dose to another group of animals which had, 20 to 30 min prior, been treated similarly with pyridoxal hydrochloride. While all rats treated only with β-cyanoalanine perished, only 13% of the jointly treated animals died. In a second experiment, pyridoxal hydrochloride was given shortly after the onset of hyperactivity; this procedure fully protected such treated rats (Ressler *et al.*, 1964). LD_{50} data obtained in an independent study that support this point constitute Table 2.

These workers also observed that subtoxic levels of β-cyanoalanine produce cystathioninuria—a condition characterized by urinary excretion of cystathionine. Since this nitrile is an effective inhibitor of rat liver cystathionase, a B_6-containing enzyme, the conversion of methionine to cysteine via cystathionine is impeded and the latter accumulates (Pfeffer and Ressler, 1967). The extent of this accumulation is exemplified by the fact that a rat receiving β-cyanoalanine for a month will excrete an average of 1.2 gm in a 24 hr period—twice that of a 40 kg child having an inborn error of metabolism preventing cystathionine cleavage (Ressler, 1975).

d. Metabolism

The published experiments of Blumenthal-Goldschmidt *et al.* (1963) were the first establishing that $H^{14}CN$ was incorporated almost exclusively into the amide-containing carbon group of asparagine (Fig. 15). This unexpected finding actually resulted from their effort to determine the contribution of free cyanide to the biosynthesis of the nitrile group of cyanogenic glycosides. Even sorghum, a monocot storing high levels of the well-known cyanogenic glycoside dhurrin, incorporated the radioactive carbon overwhelmingly into asparagine. This was true for other plants also known to contain appreciable cyanogenic glycosides, e.g., flax and white clover. In addition, serine or a related compound was implicated as the carbon source for asparagine production (Blumenthal-Goldschmidt *et al.*, 1963).

Noncyanogenic plants such as barley, pea, and red clover also synthesized considerable labeled asparagine from $H^{14}CN$. Common vetch (*Vicia sativa*) was unique in its failure to produce this amide-containing amino acid from HCN; this observation was explained subsequently by the finding that it alternatively produced γ-glutamyl-β-cyanoalanine (Ressler *et al.*, 1969).

An extract of *Lotus tenuis* catalyzed an ATP-free but serine-utilizing reaction in which 96% of the radioactivity transferred from $K^{14}CN$ to β-cyanoalanine was located in the C-4 atom (Floss *et al.*, 1965). Formation of this amino acid was enhanced 50-fold when cysteine replaced

Table 2 Effect of Pyridoxal-Hydrochloride on the LD$_{50}$ of β-Cyanoalanine[a]

Treatment[b]	Body weight		No. of animals treated	LD$_{50}$ (mg/100 gm body wt)	Ratio LD$_{50}$/LD$_{50}$ BCNA alone[c]
	Range	Average			
β-Cyanoalanine	32–41	36	20	13.4 (12.6–14.3)	—
Pyridoxal[a] plus β-cyanoalanine 20–40 min later	33–39	36	19	18.9 (16.7–21.4)	1.4 (1.3–1.6)
Pyridoxal[a] plus β-cyanoalanine 25 min later plus pyridoxal[a,e] 1¼–2½ hr later	32–37	35	19	22.5 (21.6–23.4)	1.7 (1.6–1.8)

[a] Reproduced with permission from the research of Ressler *et al.* (1967).
[b] The dosage range of β-cyanoalanine (BCNA) was 11–25 mg/100 gm.
[c] Confidence limits (19/20) are given in parentheses.
[d] Administered as pyridoxal-hydrochloride 22 mg/100 gm body weight.
[e] The second dose was given at the onset of hyperactivity.

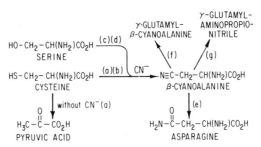

Fig. 15. Postulated metabolic relationships in higher plants involving cyanide, β-cyanoalanine, and allied compounds. Biological sources: (a) *Acacia georginae* (Mead and Segal, 1973); (b) *Lotus tenuis* (Floss *et al.*, 1965) and *Lupinus augustifolia, Vicia sativa,* and *Sorghum vulgare* (Blumenthal *et al.*, 1968); (c) *Vicia sativa* (Ressler *et al.*, 1969); (d) *Sorghum vulgare* (Blumenthal-Goldschmidt *et al.*, 1963); (e) *Lupinus angustifolius* (Lever and Butler, 1971b; Castric *et al.*, 1972) (this may be a detoxification mechanism rather than an effective *in vivo* means for asparagine biosynthesis); (f) *Vicia* (Fowden and Bell, 1965); (g) *Lathyrus odoratus* (Tschiersch, 1964).

serine, thereby implicating this sulfhydryl-containing amino acid in β-cyanoalanine biosynthesis. The efficacy of cysteine as compared to serine as a 3-carbon fragment donor for asparagine synthesis had also been reported by Blumenthal *et al.* (1968) in their study of $H^{14}CN$ assimilation into asparagine by various higher plants.

Once β-cyanoalanine is produced, it can be converted to asparagine. Castric *et al.* (1972) isolated β-cyanoalanine hydrolase activity from *Sorghum vulgare* and the blue lupin, *Lupinus angustifolius*. These workers observed that neither of these vascular plants possessed detectable asparaginase, a critical component in the eventual mobilization of the stored nitrogen of asparagine. The work of Lever and Butler (1971a) which established aspartic acid as the preferred precursor for asparagine production added further ambivalence to the question of asparagine biosynthesis. Etiolated *L. angustifolius,* which accumulated 30% of its dry weight as asparagine, was employed in their companion paper as a model system for evaluating asparagine biosynthesis (Lever and Butler, 1971b). These workers confirmed that $H^{14}CN$ effectively labeled β-cyanoalanine but a marked lag was observed between buildup of plant asparagine levels and the ontogenetic emergence of β-cyanoalanine synthase activity (Fig. 16). The ability of higher plants to sequester the carbon atom of cyanide into asparagine has also been demonstrated in *Zea mays* roots (Oaks and Johnson, 1972). In this plant, facile *in vivo* asparagine production from cyanide was discounted due to a paucity of available cyanide and the sluggishness of serine conversion to cysteine (only the latter amino acid provides carbon skeleton for asparagine production in this monocotyledenous plant).

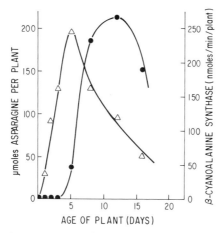

Fig. 16. Ontogenetic pattern in etiolated seedlings of *Lupinus angustifolius*. The asparagine level (triangles) and the β-cyanoalanine synthase activity (circles) are shown as a function of growth. Reprinted with permission from Lever and Butler (1971b). Copyright by Oxford University Press.

An interesting corollary to the above findings developed from the disclosure that various species of *Vicia* employ a reaction pathway culminating in the formation not of the free amino acid nitrile, but rather its γ-glutamyl peptide (Fowden and Bell, 1965). Consistent with earlier findings, certain *Vicia* species accumulate the cyanide carbon in asparagine (Table 3). These findings created the following dichotomy in reaction flow:

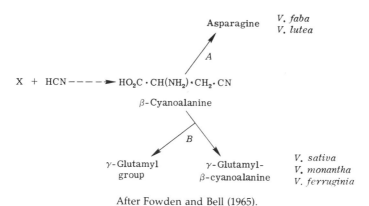

After Fowden and Bell (1965).

A sequential carbon flow from cyanide to the nitrile and then its γ-glutamyl derivative is also consistent with the data obtained from *Chlorella pyrenoidosa* (Fowden and Bell, 1965).

Table 3   [^{14}C]Cyanide Assimilation by Seedlings[a]

Species	Asparagine	γ-Glutamyl-β-cyanoalanine	β-Cyanoalanine	Aspartic acid	Other substances
Vicia sativa	0	100	0	0	0
Vicia monantha	0.8	92.6	4.5	0	2.3
Vicia ferruginia	0.3	95.2	2.3	0	2.2
Vicia lutea	89.4	1.6	0.5	5.1	3.4
Vicia faba	92.4	2.3	0.4	0.3	4.6
Lathyrus odoratus	99.1	0	0.5	0	0.4
Ecballium elaterium	23.8	19.3	1.7	1.0	54.2
Cucumis sativus	65.5	1.4	4.4	2.7	26.0

[a] Distribution of radioactivity in compounds is expressed as percentages of the total ^{14}C incorporated into 75% ethanol-soluble organic compounds. Reproduced with permission from the work of Fowden and Bell (1965).

Tschiersch (1964) has also contributed to our understanding of these metabolic reactions. His feeding studies with *Lathyrus odoratus* demonstrated appreciable incorporation of the radioactivity from H^{14}CN into the aminopropionitrile moiety of γ-glutamyl-β-aminopropionitrile. These results are in accord with the expected diversion, in lathyrogenic species, of radioactive carbon from β-cyanoalanine for the production of the lathyrogenic factor.

Tschiersch's work with *Vicia faba* suggests the existence of an assimilatory pathway in which radioactive carbon from cyanide is transferred primarily to β-cyanoalanine and a compound designated X_1. In short-term experiments, labeled X_1 does not appear. This finding is easily rationalized, as Tschiersch soon realized, if X_1 is γ-glutamyl-β-cyanoalanine. If correct this would require the relocation of *V. faba* in the above reaction flow of Fowden and Bell (1965).

The findings of these independent studies reveal that higher plants can funnel cyanide into asparagine via β-cyanoalanine. However, it is doubtful if this reaction represents a significant *in vivo* means for asparagine biosynthesis since a cyanide source is not sufficiently prevalent in higher plants. It is much more probable that β-cyanoalanine production represents an effective means for innocuous conversion of cyanide generated from such secondary plant metabolites as cyanogenic glycosides and that as such this reaction is an integral part of higher plant cyanide detoxification. Asparagine is formed from β-cyanoalanine but this may be of secondary importance to cyanide detoxification.

2. N^β-Oxalyl-L-α,β-diaminopropionic Acid

a. Background and Isolation

As mentioned previously, the correlation between neurolathyrism and human consumption of *Lathyrus sativus,* particularly under famine conditions, has been appreciated for some time. It is not surprising that this plant was the object of considerable experimental activity and that searches for its neurotoxic factor(s) culminated eventually in the isolation of the highly potent neurotoxic natural product: N^β-oxalyl-α,β-diaminopropionic acid (ODAP).

$$HOOC-C(=O)-NH-CH_2-CH(NH_2)COOH$$

N^β-Oxalyl-L-α,β-diaminopropionic acid

As early as 1964, Rao *et al.* isolated this oxalyl derivative from *L. sativus* seed; its identity was confirmed by a chemical synthesis in which dimethyl oxalate was employed for oxalylating α,β-diaminopropionic acid. This neurotoxin also occurs in *L. cicera* and *L. clymenum* (Bell and O'Donovan, 1966) as well as in *Crotalaria incana* and *C. mucronata* (Bell, 1968). In a recent and extensive survey of 250 plant genera, the oxalyl derivative was demonstrated in 13 species of *Crotalaria,* 17 members of *Acacia,* and 21 *Lathyrus* species (Qureshi *et al.,* 1977). Values on the extent of its storage in *L. sativus* seeds vary from 0.1 to 2.5% (dry weight basis) (Roy and Narasinga Rao, 1968). Another neurotoxin has been isolated from the above *Lathyrus* species, namely, N^β-D-glucopyranosyl N-arabinosyl-α,β-diaminopropionitrile. A dose of 50 mg/100 gm body weight produces limb paralysis within 5 to 10 min (Rukmini, 1969).

b. Biological Properties and Toxicity

When ODAP is administered to an adult mammal so that the blood–brain barrier is breached, head retractions, tremors, and convulsions result (Rao *et al.,* 1967). It is believed that this barrier is not yet developed in young animals; thus, while intraperitoneally injected ODAP (5 to 8 mg) causes convulsive seizures in young rats within 5 min that occasionally prove fatal, adult rats receiving three times this dosage are not affected (Rao and Sarma, 1967). On the other hand, induction of an "acidotic" state overcomes the blood–brain barrier toward ODAP (Rao and Sarma, 1967). As a result, when acidotic adult rats received 2.2 mmoles ODAP per kg body weight, they experience convulsive seizures and brain urea,[*]

---

[*]An earlier observation by these workers (Cheema *et al.,* 1969b) disclosed that ODAP does not affect urea in the brain of young rats. It is possible, of course, that acidosis itself is responsible for the urea noted in acidotic adult rats.

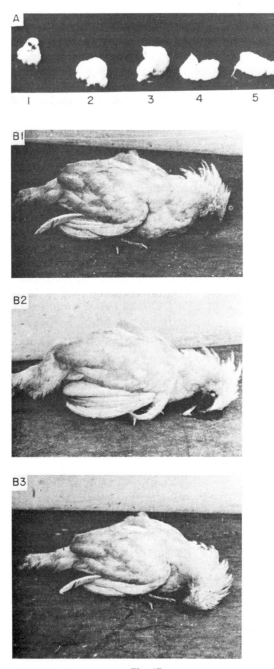

Fig. 17.

ODAP, ammonia as well as glutamine levels increase markedly (Cheema *et al.*, 1969a).

If humans have an effective blood–brain barrier toward ODAP, it would represent a fortuitous property rendering protection to adults consuming *L. sativus*. One must wonder, however, if poor nutrition mitigates physiological conditions eroding this protective barrier since the adverse biological effects of ODAP are intensified greatly during famine conditions. Other factors may operate during the stress of food deprivation which are unrelated entirely to this protective barrier. Finally, Cheema *et al.* (1969a,b) postulated that the overall toxicity of this oxalyl-containing compound results from elevated brain ammonia that potentiates convulsive seizures. Ammonia detoxification in the brain occurs normally by sequestration as the amide nitrogen of glutamine but this ammonia sink can be saturated readily.

This oxalyl derivative is also very toxic to avians; injections of 20 mg per chick caused wry neck, head retraction, and convulsions (Adiga *et al.*, 1963). Administration of a lesser dose, i.e., 10 mg per chick, produces less severe symptomology while 30 mg per chick is lethal. Nargarajan *et al.* (1965) reported neurological aberrations and opisthotonus in ducklings and baby pigeons given 1.4 mg ODAP per gm body weight. Adult birds do not respond to this toxic metabolite when it is administered intraperitoneally, but are susceptible if rendered acidotic with $CaCl_2$ (Fig. 17).

The capacity of various nonprotein amino acids to function as nerve cell excitants has been evaluated by Watkins *et al.* (1966) by recording extracellularly the action potentials from spinal interneurones and Betz cells. Over a pH range of 2.8 to 3.4, β-cyanoalanine neither excited nor depressed these test cells. α,γ-Diaminobutyric acid attenuated synaptic responses but much less so than γ-aminobutyric acid or β-alanine. Their results suggest a mode of action other than direct nerve cell action for the above two neurotoxic plant constituents. Similar experiments with ODAP administered at pH 7.5 reveal that it is a highly efficacious excitant. Indeed, it is only slightly less active than N-methyl-D-aspartic acid, the strongest excitant of feline nerve cells known at the time of their study (Watkins *et al.*, 1966). In contrast, the oxalyl derivative of α,γ-

---

Fig. 17. N^β-Oxalyl-α,β-diaminopropionic acid (ODAP) and the chick. Chick 1 in (A) is the control organism, while chicks 2–5 illustrate the progressive symptomology observed in this avian. The adult bird shown in (B1)–(B3) was rendered acidotic with calcium chloride prior to ODAP treatment by intraperitoneal injection. After Serma and Padmanaban (1969).

diaminobutyric acid did not perceptively stimulate these test cells. A recent analysis of several parameters of brain tissues secured from ODAP-treated young rats indicated a biochemical pattern in conformity with that induced by an excitant amino acid (Cheema et al., 1970).

In his assessment of the biological action of ODAP, Johnston (1973) proposed that it may mimic the postsynaptic action of excitatory transmitters functioning in the central nervous system and that its ability to induce convulsions may reflect its direct excitant action on central neurons. While this supposition is entirely reasonable, experimental evidence for such action by ODAP in the central nervous system has not been presented.

Other amino acids are capable of producing neurotoxic effects. The D and L racemate of glutamic acid, the enantiomers of aspartic acid, and cysteine are known neurotoxins. Administration of glutamic acid as its monosodium salt (MSG) to neonates produces adults that can be stunted and obese; possess anomalous estrous cycling; exhibit evidence of infertility; and atropy of the pituitaries, ovaries, and uteri (Kizer et al., 1978). It is little wonder that this food flavor enhancer has been banned from baby foods but it remains an overused component of some adult diets.

In his review article of neurotoxic amino acids, Johnston (1973) generalized that numerous monoamino, dicarboxylic acids (or their equivalent) are "excitants," i.e., capable of exciting locally treated neurons. Many of these amino acids produce convulsions in adults and cause brain neuron degeneration of immature mammals. Compounds such as γ-aminobutyric acid, in contrast, represent a group that inhibit transmission of impulses in the central nervous system. These amino acids are potent inhibitors at the synapses of neurons and skeletal muscular fibers of arthropods and are considered a major inhibitory transmitter of the human central nervous system.

N^β-Oxalyl-α,β-diaminopropionic acid is currently the most potent of the higher plant neurotoxic amino acids but it is much less active neurologically than several heterocyclic isoxazole analogs associated with fungi. These constituents include kainic acid, quisqualic acid, and domoic acid. Two other neurologically active nonprotein amino acids of fungi are ibotenic and tricholomic acids (see Johnston, 1973).

In Lathyrus sativus, [U-^{14}C]oxalic acid is incorporated intact into N^β-oxalyl-α,β-diaminopropionic acid (Malathi et al., 1967) in a reaction sequence believed to be the composite of two integrated reactions (Malathi et al., 1970):

$$\text{Oxalate} + \text{ATP} + \text{coenzyme A} \xrightarrow[\substack{\text{oxalyl-CoA} \\ \text{synthetase}}]{\text{Mg}^{2+}} \text{oxalyl-CoA} + \text{AMP} + \text{PP}_i$$

$$\text{Oxalyl-CoA} + \text{L-}\alpha,\beta\text{-diaminopropionic acid} \xrightarrow[\substack{\text{ODAP} \\ \text{synthase}}]{} \text{ODAP} + \text{coenzyme A}$$

An oxalyl-CoA synthetase has been purified partially from *Pisum sativum* (Giovanelli, 1966) but ODAP is not a natural product of this legume. The substrate specificity of the amino acid acceptor for these leguminous enzymes has not been determined fully, but it does appear to be somewhat broad. Oxalyl derivatives of alanine, glycine, serine, homoserine, and lysine are formed, and in *L. sativus*, the α-amino and hydroxyl groups of amino acid acceptors can be oxalylated (Johnston and Lloyd, 1967). Relative product formation with various substrates by the oxalyl-generating system of *L. sativus* is revealed by the data of Table 4. Malathi *et al.* (1970) reached the following conclusions: (a) a 3-carbon amino acid represented the optimum substrate for oxalylation; (b) this 3-carbon amino acid's β-amino group exhibits a maximum reaction rate providing the α-carbon also has an amino group; (c) in hydroxyamino acids, the γ-hydroxyl is more reactive to oxalic acid than the β-hydroxyl

Table 4 Oxalylation of Amino Acid Acceptors[a]

Amino acid acceptor	Oxalyl derivatives (μmoles)	
α,β-Diaminopropionic acid	β-N-Oxalyl derivative	3.10
	dioxalyl derivative	1.90
α,γ-Diaminobutyric acid	γ-N-Oxalyl derivative	0.06
	dioxalyl derivative	0.90
Homoserine	O-Oxalyl derivative	0.30
	N-oxalyl derivative	0.49
	dioxalyl derivative	0.21
Serine	O-Oxalyl derivative	0.02
	N-oxalyl derivative	0.60
	dioxalyl derivative	0.20
Glycine	N-Oxalyl derivative	0.20
Alanine	N-Oxalyl derivative	0.13
β-Alanine	N-Oxalyl derivative	0.15
α-Aminobutyric acid	N-Oxalyl derivative	0.19
γ-Aminobutyric acid	N-Oxalyl derivative	0.12

[a] Reproduced with permission from the research of Malathi *et al.* (1970).

group. These workers also proposed that the reactivity of various functional groups in enzyme-mediated oxalylation can be considered to obey the following order: β-amino group of α,β-diaminopropionic acid > γ-amino group of α,γ-diaminobutyric acid > α-amino group of homoserine or serine > γ-hydroxyl group of homoserine > β-hydroxyl group of serine (Malathi *et al.*, 1970).

3. L-α,γ-Diaminobutyric Acid and Its N^γ-Oxalyl Derivative

a. Background and Isolation

While searching for the toxic constituent of *Lathyrus sylvestris* and *L. latifolius*, Ressler *et al.* (1961) achieved the first isolation of α,γ-diamino-butyric acid (DABA).

$$H_2N—CH_2—CH_2—CH(NH_2)COOH$$

L-α,γ-Diaminobutyric acid

Seeds of *L. sylvestris* W. contain about 1.6% DABA by dry weight but in terms of total nitrogen, it accounts for $10.3 \pm 0.09\%$ of the seed's supply of nitrogen (van Etten and Miller (1963). This toxic diamino acid was not found by Bell (1962a) in such harmful seeds as *L. sativus, L. cicera,* or *L. clymenum,* but van Etten and Miller (1963) isolated DABA from *L. cicera* samples. This conflicting finding may result from a natural concentration range resting close to the level of detectability. At least 11 other species of this genus also store DABA, typically constituting about 1% of the seed's dry weight (Bell, 1962b). It is evidently much less abundant in other genera since an extensive survey, covering 145 species of the Leguminosae and Compositae, revealed a consistent pattern of only trace-level occurrence (van Etten and Miller, 1963).

High-voltage electrophoretic analysis of a seed extract prepared from *L. latifolius* separated a N^γ-oxalyl derivative which in this particular sample was even more abundant than the parent compound-DABA. Hydrolysis of N^γ-oxalyl-α,γ-diaminobutyric acid produced the expected reaction products: oxalic acid and DABA (Bell and O'Donovan, 1966). These workers also discovered a temperature-dependent equilibrium reaction between the N^α- and N^γ-oxalyl derivatives; the occurrence of an unstable reaction intermediate was hypothesized:

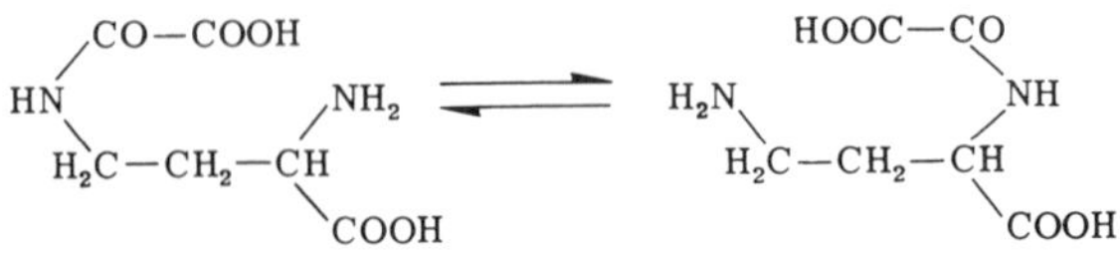

$$\begin{array}{c} \text{CO} \underline{\quad\quad} \text{CO} \\ \diagup \quad\quad\quad \diagdown \\ \text{HN} \quad\quad\quad\quad \text{NH} \\ \diagdown \quad\quad\quad \diagup \\ \text{H}_2\text{C} - \text{CH}_2 - \text{CH} \\ \diagdown \\ \text{COOH} \end{array}$$

(Possible intermediate)

In 10 of 13 *Lathyrus* species that produced DABA, both the N^α-oxalyl and N^γ-oxalyl derivative also occur; the latter compound is more abundant.

b. Biological Effects and Toxicity

Stomach-tube intake of this compound by male weanling Sherman rats produces a weakness in the hind legs within 48 hr. Upper extremity tremors occur abruptly prior to the onset of a convulsive state which preceeds death (Ressler *et al.*, 1961). DABA concentrates preferentially in the liver of this mammal (Christensen *et al.*, 1952); significant impairment of liver function would elevate blood and brain ammonia levels and initiate convulsive seizures. This factor was considered by O'Neal *et al.* (1968) in their thorough investigation of DABA toxicity in male albino rats. These workers noted that intraperitoneal injection of 4.4 mmoles DABA per kg body weight caused hyperirritability, tremors, and convulsions in 12 to 20 hr; death ensued 3 to 8 days later. After a 30 hr treatment, blood and brain ammonia levels for treated vs. control animals were 0.18 ± 0.039 vs. 0.11 ± 0.018 μmoles ml^{-1} and 1.70 ± 0.16 vs. 1.32 ± 0.10 μmoles gm^{-1}, respectively; treated rats had more brain glutamine.

The most frequently quoted of their observations is the ability of DABA to curtail albino rat ornithine carbamoyltransferase; comparable findings were reported with this enzyme from *Neurospora crassa* (Herrmann *et al.*, 1966). At the same time, the treated rats exhibited elevated blood urea. Inhibition of ornithine carbamoyltransferase would reduce the ornithine available for urea formation. This apparent contradiction could be rationalized, however, in terms of DABA-directed diuresis which could increase the urea concentration while urea production actually diminished.

DABA-treated rats were characterized as exhibiting only chronic ammonia toxicity in that ammonia production progressed at a slow rate. This is in contrast to the acute toxicity induced by other amino acids or ammonia (O'Neal *et al.*, 1968). Chicks treated with nearly three times the DABA used in the above experiments with rats evinced no discernible symptoms. Such resistance by the uricotelic chick as compared to the ureotelic rat may result from the lack of avian dependency on urea production for nitrogen excretion. This finding is consistent with the involvement of DABA in attenuating urea production while enhancing internal ammonia as a significant basis for its biological toxicity.

In a study of nonprotein amino acid/insect interaction conducted with the red flour beetle, *Tribolium castaneum*, Applebaum and Schlesinger (1977) characterized DABA as being only somewhat toxic, much less so than β-cyanoalanine. Larvae of the yellow mealworm, *Tenebrio molitor*, are appreciably more sensitive to both of these neurotoxic compounds (Fig. 18), and their sensitivity to β-cyanoalanine relative to DABA is much greater. These findings with *Tenebrio* do not reflect merely a lack of larval feeding activity, for these organisms are able to sustain a longer survival period even when starved. *Tenebrio* larval fresh weight actually increases during the larval instar (1% DABA-supplemented diet) but the terminal stadium larvae perished ultimately since they cannot ecdyse or metamorphose to pupae.

c. Metabolism

Information on the metabolism of DABA is limited but there is some experimental evidence for its formation from β-cyanoalanine (See Fig. 13, p. 61). However, administration of L-[4-^{14}C]β-cyanoalanine to *Lathyrus sylvestris* culminated in the transfer of only 0.27% of the label to

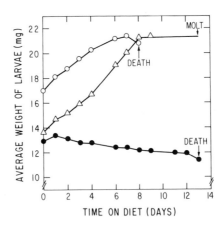

Fig. 18. The effect of α, γ-diaminobutyric acid and β-cyanoalanine on the fresh weight development of the yellow mealworm, *Tenebrio molitor*. Larval diet was supplemented with 1% diaminobutyric acid (open circles) or β-cyanoalanine (solid circles); unsupplemented diet served as the control (triangles). After Applebaum and Schlesinger (1977).

α,γ-diaminobutyric acid (Nigam and Ressler, 1966). Tracer feeding studies with L-[^{3}H]homoserine and DL-[1-^{14}C]aspartic acid provided much more substantive evidence to implicate these compounds in DABA synthesis in this legume.

DABA has been proposed as an intermediate in the formation of azetidine-2-carboxylic acid from homoserine (Fig. 32, p. 138) and it is believed to function in the biosynthesis of N^γ-diaminobutyric acid, employing an oxalyl-activating system using ATP·Mg and coenzyme A that parallels the production of N^β-oxalyl-α,β-diaminopropionic acid (see p. 61). The observation of Nigam and Ressler (1966) that the labeled carbon transferred from homoserine and aspartic acid to DABA failed to label effectively its N^γ-oxalyl derivative is most intriguing.

A report of the toxicity of this derivative is limited to the observation of Rao and Sarma (1967) that it is neurotoxic to 1-day-old chicks. Obviously, much less is known of this oxalyl-containing natural product than is known of its lower homolog.

4. N^β-Methyl-L-α,β-diaminopropionic Acid

The Cycadales, comprised of nine extant genera, were a dominant vegetation of Mesozoic Earth. These "palmlike" plants are indigenous to tropical and subtropical floristic regions where they are consumed both as an article of human diet and by domesticated animals (Vega and Bell, 1967). Cycad-eating cattle are susceptible to a neurological disorder in which the hindquarters are paralyzed reversibly (Mason and Whiting, 1966). This report instigated the search by Bell and associates for the toxic principle of this unusual plant group—apparently a β-methylated derivation of α,β-diaminopropionic acid.

$$H_3C—NH—CH_2—CH(NH_2)COOH$$

N^β-Methyl-L-α,β-diaminopropionic acid

Initially obtained from the seeds of *Cycas circinalis* (Vega and Bell, 1967), eight other species of *Cycas* also store this compound (Dossaji and Bell, 1973). α,β-Diaminopropionic acid has been sought for but not obtained in these plants. The β-methyl derivative content in the leaves and seeds (less than 0.02% fresh weight) is insufficient to account for their observed toxicity. This point was appreciated earlier since Bell and associates had determined that long-term chronic exposure could not elicit the observed symptomology. For this determination, a racemic mixture (0.28 mg or 2.4 μmoles per gm) was administered subcutaneously to rats for 78 consecutive days; while this treatment failed to produce significant neurological effects, overall growth was adversely affected (Polsky *et al.*, 1972).

In contrast, a single intraperitoneal injection (0.84 mg or 7.2 μmoles per gm body weight) induced convulsions and produced considerable rat mortality. Chicks appeared somewhat more sensitive and mice somewhat less sensitive. Subsequent analysis by Vega *et al.* (1968) revealed that the D enantiomer, in both the chick and rat, lacked discernible biological activity. Thus, while the toxicity of this natural product has been established, by itself it cannot account for the human neurological disease resulting from consumption of these plants. Definitive information is not available on the metabolism of this compound, but it can form by simple methylation of α,β-diaminopropionic acid.

B. HETEROCYCLIC AND SUBSTITUTED AROMATIC COMPOUNDS

1. L-Mimosine

a. Background and Isolation

Mimosa pudica and *Leucaena leucocephala* (previously *L. glauca* Benth.) represent the initial plant sources of L-mimosine, a toxic heterocyclic nonprotein amino acid having the structure β-[N-hydroxy-4-oxypyridyl]-α-aminopropionic acid.

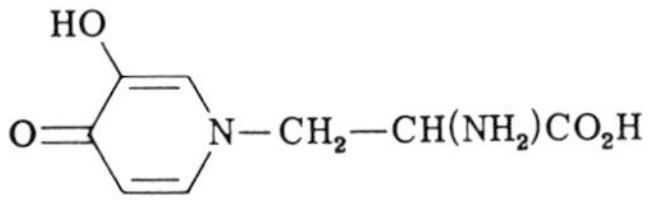

L-Mimosine

On a dry weight basis, mimosine constitutes about 2 to 5% of the leaf of *L. leucocephala* (Gonzalez *et al.*, 1967; Brewbaker and Hylin, 1965) and as much as 9% of the seed (Takahashi and Ripperton, 1949). Mimosine was named by Renz (1936) who obtained it from *M. pudica* sap. It was the chemical studies of Adams *et al.* and Bickel (quoted in Suda, 1960) that established that leucenol, also known as leucaenol or leucaenine, was mimosine. Thus, a single toxic nonprotein amino acid exists and it is limited presently to the above two genera of the Mimosaceae.

Leucaena leucocephala, an arborescent legume, is indigenous to Mexico but it has been planted widely throughout the neotropics and tropics where it serves as a windbreak and cover for such crops as coffee, tea, and cocoa. It is an aggressive species, especially on xerophytic sites (Brewbaker and Hylin, 1965) and can reach heights of 60 feet. Its ability to thrive on marginal soils, to withstand repeated defoliation, its marked productivity, palatability, as well as its nitrogen fixing capability* and crude protein and nitrogen contents have all combined to make it a valuable wood product and fodder crop.

b. Biological Properties

The high potential of *L. leucocephala* as a commercial crop has stimulated many higher animal studies of the toxic principle sequestered in the leaf and seed. Such investigations established that contrary to certain reports of its innocuous nature in dairy and beef cattle (e.g., Henke, 1959), grossly enlarged thyroid glands are formed by calves produced from heifers consuming this legume (Hamilton *et al.*, 1968). Both Jones *et al.* (1976) and Hegarty *et al.* (1976) have confirmed this observation. They also showed that a metabolic derivative, namely 3-hydroxy-4-(1-*H*)-pyridinone[†] is responsible for the goitrogenic condition (Hegarty *et al.*, 1979). While this legume has well documented forage qualities, its value for monogastric animals is compromised by weight loss, a general malaise, eye inflammation, and marked depilation (Joshi, 1968). These limitations are not applicable to ruminants since they harbor a symbiotic population capable of detoxifying mimosine.

In a 1905 report on the Bahamas prepared by the Baltimore Geographic Society it is stated: "We also found in these islands specimens of the Jumbie Bean (*L. leucocephala*). This bean, when eaten by mules or horses, causes the hair of the mane and tail to fall out, giving a rather remarkable appearance to the animals who have been unfortunate enough to have the bean as a diet" (Owen, 1958).

Owen (1958) confirmed these observations and noted in describing his own work that when a horse consumes mimosine, the mane and tail hairs depilate—sometimes with such severity that little hair remains. This effect can be reversed by transferring the inflicted animal to mimosine-free forage. Swine are appreciably more susceptible since mimosine consumption can render this ungulate completely bald. Apparently, goats (Kraneveld and Djaenoedin, 1950) and beef and dairy

---

*The nitrogen fixation rate has been estimated at about 600 kg/hectare/year.

†This is the tautomeric keto form of 3,4-dihydroxypyridine and probably represents the natural product species.

cattle (Hutton and Gray, 1959) are generally immune to this type of toxic effect.

In their investigation of purified mimosine and ground L. *leucocephala* seed and mice, Crounse *et al.* (1962) plucked hair from a large body area to stimulate new hair follicle growth. Thick coats appeared in depleted areas of the control animals within 8 to 10 days and even mice that consumed a diet containing 5% ground seed of 0.5% mimosine were able to grow hair normally. No new pubescence developed, however, when the ground seed was at 10% or the mimosine level at 1.0%.

Mimosine does not affect resting hairs; damage is limited consequently to periods of active hair growth. Any hair loss must therefore reflect the extent to which the root of the *growing* hair is damaged. Some hair loss is typical, but since mimosine blocks normal hair replacement, the total hair content gradually diminishes (Crounse *et al.*, 1962). The hair growth phase is of a much shorter duration than the mimosine-resistant resting phase; this factor may account for the contradictory reports of mimosine's depilatory properties—particularly in regard to dairy beef and cattle (Hegarty *et al.*, 1964). Hylin's 1969 report that mimosine is a strong inhibitor of rat liver cystathionine synthetase and cystathionase is germane, for these enzymes convert methionine to cysteine via cystathionine. Cysteine is an important component of hair follicles and the failure of epilated areas to regain their original hirsute state may be related to decreased cysteine production from methionine.

Mimosine's depilatory capacity has been capitalized on as a sheep defleecing agent (Fig. 19). The amount, nature, and duration of the mimosine administration affect various aspects of this process. For example, at a level of 300 mg per kg body weight, two successive oral doses constituted an effective defleecing regime but a single oral dose of 450 to 600 mg per kg achieved the same end (Reis *et al.*, 1975b). Continuous intravenous infusions (totaling 4 gm over a period of 1.5 to 2 days) defleeced the treated sheep while twice this amount, given by intraperitoneal injection, was ineffective (Reis, 1975; Reis *et al.*, 1975a).

If the dose is relatively low, such as 1 gm per day, but given over an extended period, marked diminution in wool fiber diameter and strength is discernible after 10 treatment days (Frenkel *et al.*, 1975). One can readily appreciate that if the dose rate and method of application is controlled properly, Reis *et al.* (1975a) are justified in stating: "It is apparent that mimosine has a potential use as a chemical defleecing agent if a practical means of administration could be found to obtain a controlled rate of release into the bloodstream." The work of these Austra-

Fig. 19. Sheep defleecing. This sheep, provided 4 gm mimosine/day for 2 days by intravenous infusion, was defleeced chemically 15 days after initiation of treatment. Reprinted with permission from Reis *et al.* (1975b). Copyright by CSIRO. Photo provided by P. J. Reis.

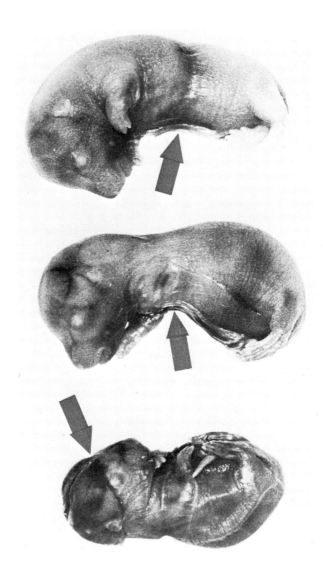

Fig. 20.   Rat uterine wall perforations. The illustrated deformities are caused by fetal protrusion through the uterine perforations. The arrows indicate regions constricted by the uterine tissue surrounding perforations. Reprinted with permission from Dewreede and Wayman (1970). Copyright by Wistar Institute Press. Photo provided by the authors.

lian investigators has resulted in a novel and potentially useful application of an otherwise toxic plant constituent. This is particularly true if
this compound degrades rapidly, thereby avoiding a toxic residue that
can adversely affect its utility in meat production.

Maintaining female Sprague–Dawley rats on a prolonged regime of
0.5% mimosine-containing diet produces irregular estrous cycling. If the
mimosine dietary content is doubled, the estrous cycle terminates (Hylin
and Lichton, 1965). Overall reproduction is also disrupted, as manifested by small litter size in rabbits (Willet *et al.,* 1947) and sows
(Wayman and Iwanaga, 1957), reduced egg production in poultry
(Thanjan, 1967), and mouse embryo fatality (Bindon and Lamond,
1966).

Rats fed a diet supplemented with 0.5 to 0.7% mimosine produced
fetal offspring with deformed cranium, thorax, and pelvis as well as
sacral vertebrae that failed to develop. The deformities were associated
with uteral wall perforations (Fig. 20) that led Dewreede and Wayman
(1970) to report: "The perforations were located near the placental at-
tachment, and the uterine wall surrounding the perforation was thick-
ened around the edges. Parts of fetuses, still in the fetal membranes,
protruded through the uterine perforations. It was apparent that deformities were the result of constriction of the protruding fetal parts by
uterine tissues."

c. Toxicity

The hydroxyproline residues of collagen result from hydroxylation of
the proline constituents of protocollagen, the progenitor of collagen
(Peterkofsky and Udenfriend, 1963). Mimosine inhibits proline hydroxylation and intracellular protocollagen accumulates although [^{14}C]proline
continued to be incorporated into protein (Tang and Ling, 1975).
Hydroxyproline-deficient collagen is more susceptible to the degradative action of collagenase (Hurych *et al.,* 1967). Thus, any reduction in
total collagen synthesis resulting from impeded hydroxyproline formation produces a more ephemeral collagen and this can ramify ultimately
into capillary hemorrhaging of the uterine perforations observed in
certain mimosine-fed animals and shown in Fig. 20 (Tang and Ling,
1975).

An appreciation has emerged of other modes of action at the biochemical level. This heterocyclic compound, as with canaline, reacts
with the aldehydic carbon atom of pyridoxal phosphate (Lin *et al.,*
1962a).

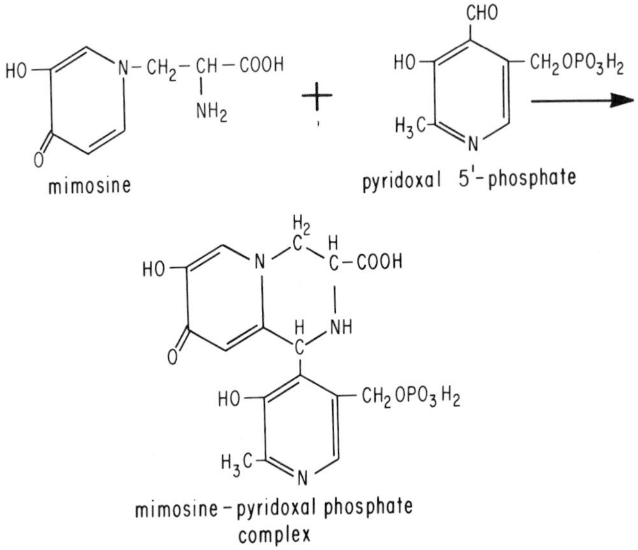

mimosine   pyridoxal 5'-phosphate

mimosine – pyridoxal phosphate
complex

This reaction, sequestering pyridoxal phosphate, may disrupt the catalytic action of B_6-containing enzymes, and mimosine-mediated inhibition of pyridoxal phosphate-dependent glutamic-aspartic transaminase of swine heart (Lin *et al.*, 1962a) and L-dopa decarboxylase of swine kidney (Lin *et al.*, 1963) have been demonstrated.

Either ferrous sulfate or dietary intake of mimosine as the ferrous salt reduces its intrinsic toxicity (Matsumoto *et al.*, 1951; Lin and Ling, 1961); the sparing effect of Fe^{2+} may result from interference in the formation of a complex between mimosine and pyridoxal phosphate (Lin *et al.*, 1962a). Mimosine may also chelate metallic ions required for enzyme activation (Tang and Ling, 1975). In this regard, Mg^{2+} assuages mimosine-dependent inhibition of alkaline phosphatase (Chang, 1960).

There is some evidence that mimosine functions as a tyrosine antimetabolite. It suppresses tyrosine decarboxylase and tyrosinase activities (Crounse *et al.*, 1962); comparable findings have been obtained by Prabhakaran *et al.* (1973). Frenkel *et al.* (1975) infused sheep (*Merino wethers*) intravenously with mimosine to test its influence on the tyrosine content of the wool and noted a reduction in both total tyrosine and the high tyrosine-containing protein content of the wool. Neither tyrosine nor phenylalanine, given at a 3-fold excess relative to a mimosine dose of 1 gm per day, counteracted mimosine's action on wool. It is possible that a higher ratio of these aromatic amino acids to mimosine may have spared the observed toxicity, particularly since Lin *et al.* (1964) reported that tyrosine and phenylalanine minimize certain adverse effects of mimosine in animals. The question of whether mimosine is a phenylalanine or tyrosine antimetabolite (or both or neither) is further obfuscated by the finding that phenylalanyl- but not tyrosyl-tRNA synthetase of mung bean activates mimosine. When mimosine was provided at 60 times the phenylalanine level, formation of phenylalanyl-tRNA was not impeded (Smith and Fowden, 1968). Thus, in mung bean, mimosine does not appear to affect significantly either phenylalanine or tyrosine placement into protein.

Another aspect of mimosine's biochemical action has been elucidated by work of Ward and Harris (1976) in which sheep skin tissue slices served as the *in vitro* experimental test system. The effect of this heterocyclic natural product on wool production was evaluated by assessing follicle function. Radioautographic analysis of skin slices treated with [³H]thymidine reveal the expected high density labeling of the follicle bulb nuclei. With mimosine, a much weaker labeling pattern ensues (Fig. 21). While mimosine inhibits effectively DNA synthesis, it has no effect on protein synthesis. Their research efforts also reveal that a racemic mixture is as potent as an equivalent amount of the

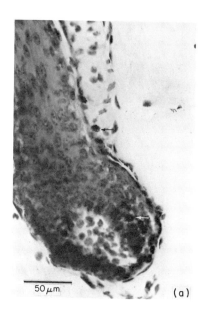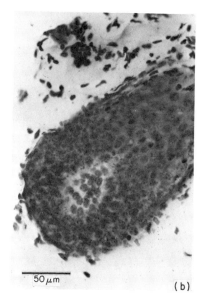

Fig. 21. Autoradiographs of [³H]thymidine-labeled skin slices. The tissues were incubated in Waymouth's medium without mimosine (a) or with 0.2 mM mimosine (b). The arrows in (a) depict heavily labeled nuclei. Reprinted with permission from Ward and Harris (1976). Copyright by CSIRO.

L-enantiomorph (Fig. 22). Ward and Harris tested an array of compounds related structurally to mimosine for their ability to curtail DNA synthesis (Table 5) and established the critical role of the 3-hydroxy-4-oxo moiety of mimosine in eliciting this biochemical effect.

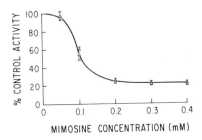

MIMOSINE CONCENTRATION (mM)

Fig. 22. Mimosine and DNA production. Each point is the mean of 12 determinations. Treatments: L-mimosine (circles) and DL-mimosine (triangles). Percent control activity refers to the level of [³H]thymidine incorporated into skin slices relative to the control. Reprinted with permission from Ward and Harris (1976). Copyright by CSIRO.

Table 5   Inhibition of Wool Follicle DNA Synthesis by Mimosine and Related 4(1H)-Pyridones[a]

Compound	Percentage of control value incorporated[b]
L-Mimosine	28 ± 2
DL-Mimosine	28 ± 3
L-Mimosine methyl ester	34 ± 5
DL-Isomimosine	27 ± 2
3-Hydroxy-4-(1H)-pyridone	35 ± 4
DL-2-Methylmimosine	28 ± 2
3-Hydroxy-2-methyl-4(1H)-pyridone	21 ± 1
Mimosinamine	74 ± 9
N,N-Dimethylmimosinamine	35 ± 5
DL-α-Hydroxydesaminomimosine	92 ± 4
DL-N-Benzoylmimosine	104 ± 10
DL-3-O-Benzyl-2-methylmimosine	100 ± 14
DL-3-Deoxymimosine	101 ± 10
DL-4-Deoxyisomimosine	100 ± 18
DL-Mercaptoisomimosine	66 ± 5

[a] Skin slices were incubated in Waymouth's medium containing 0.2 mM of the appropriate inhibitor. Reproduced with permission from the work of Ward and Harris (1976). Copyright by CSIRO.
[b] Refers to the level of [³H]thymidine incorporated into skin slices relative to control material.

d. Metabolism

Several piperidine compounds including mimosine and pipecolic acid are stored by *L. leucocephala*. Since pipecolic acid is known to be synthesized from lysine, Hylin (1964) fed DL-[2-^{14}C]lysine to this legume as part of his study of mimosine production. The subsequently isolated mimosine exhibited a high specific activity and the radiochemical yield was 3.6%. Mimosine pyrolysis yielded 95% of the incorporated radioactivity as 3,4-dihydropyridine, thereby establishing that lysine almost totally labeled the pyridone ring. These findings supported a putative biosynthetic link between lysine and mimosine.

A similar type of experiment with *Mimosa pudica* indicated that the radioactivity of [3-^{14}C]aspartic acid enters into the same pyridone moiety (Notation and Spenser, 1964). These workers hypothesized that aspartate, after loss of its C-4 atom, was incorporated into lysine prior to mimosine formation. Neither DL-[1-^{14}C] nor DL-[4-^{14}C]aspartic acid labeled the pyridone nucleus. They also presented evidence implicating the α-aminoadipic acid pathway rather than diaminopimelic acid in lysine formation (see p. 168). Tiwari *et al.* (1967) determined subsequently that serine served as the carbon source of the alanyl side chain. The extent of DL-[6-^{14}C]α-aminoadipic acid incorporation into the pyridone ring provided further credence to the putative role of this pathway in the conversion of aspartic acid to lysine prior to mimosine formation. It is relevant that δ-hydroxylysine is not incorporated into mimosine; this suggests that hydroxylation is achieved only after the pyridone ring is formed (Tiwari *et al.*, 1967).

In their study of wool abscission from sheep, Reis *et al.* (1975b) evaluated the effects of oral intake of mimosine and found it to be excreted largely unaltered after 1 day of treatment. By the second day, significant amounts of 3,4-dihydroxypyridine (DHP) appeared in the urine. Intravenous or abomasal administration failed to produce detectable DHP—thereby suggesting that this catabolic reaction was mediated by rumenal microflora. These workers suggested that the observed time lag reflected either the time required for induction of the appropriate degradative enzymes or a recovery period necessitated by the decimation of the rumen's symbiotic population. This finding confirmed an earlier study by Hegarty *et al.* (1964) in which extensive detoxification of mimosine via DHP occurred only in the rumen.

In this regard, *Leucaena* seedling extracts also catalyze a stoichiometric conversion of mimosine to DHP, pyruvate, and NH_3(Smith and Fowden, 1966). According to their growth inhibition studies, conducted with mung bean radicles, DHP exhibits a toxicity equal to mimosine. While ferrous ions alleviate the toxic effects of DHP and mimosine, only the action of mimosine is spared by pyridoxal phosphate. Thus, in this higher plant 3,4-dihydroxypyridine formation is not an effective means for detoxification of mimosine.

2. 5-Hydroxy-L-tryptophan

Hydroxylation of tryptophan to produce 5-hydroxytryptophan is recognized as the opening step in mammalian conversion of the protein amino acid to 5-hydroxytryptamine (Udenfriend *et al.*, 1956). This physiologically active amine, known as serotonin, can be converted to a major animal pigment, melatonin, by acetylation and methoxylation.

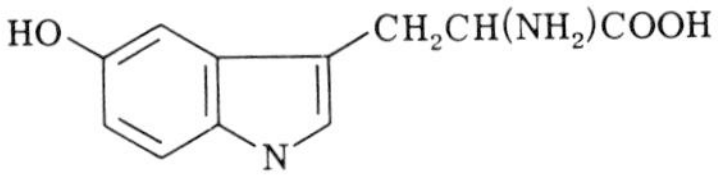

5-Hydroxy-L-tryptophan

5-Hydroxytryptophan has been obtained from the legume *Mucuna pruriens* (an important source of L-dopa) as well as banana and cotton. It was first obtained in large amounts from the West African legume *Griffonia simplicifolia* where it can account for 6 to 10% of the fresh weight of the seed. Analysis of the various tissues of *G. simplicifolia* reveal that with the exception of the mature seed, the tested tissues vigorously hydroxylate L-[3-^{14}C]tryptophan in the 5 position. After an 18 hr reaction period, 5-hydroxytryptophan accounts typically for 70% of the radioactivity detected in the treated leaf discs (Fellows and Bell, 1970).

While serotonin appears to be involved in central nervous system function, it cannot breach the blood–brain barrier; this restriction does not apply to 5-hydroxytryptophan. It is possible, therefore, to enhance significantly brain levels of the amine since it can form readily by decarboxylase activity of this organ. Elevated brain serotonin creates a behavioral syndrome reminiscent of lysergic acid diethylamide which include tremors, lack of muscular coordination, hyperventilation, accelerated heartbeat, salivation, apparent blindness, pupillary dilation, and lachrymation. These effects are noted in both dogs receiving 30 to 60 mg per body weight and rabbits administered 60 to 100 mg per kg body weight (Udenfriend *et al.*, 1957). Treating these animals with a drug blocking serotonin degradation accelerates the onset and intensity of the above symptoms and enhances overall sensitivity; more importantly, the test organisms quickly perished. Ingestion of this hydroxylated nonprotein amino acid can augment serotonin levels in both the brain and other body organs and potentiate the adverse biological effects elicited by this physiologically active amine.

3. 3,4-Dihydroxy-L-phenylalanine (L-Dopa)

In 1913, Torquati and Guggenheim independently announced the simultaneous isolation of this hydroxylated compound from *Vicia faba*. Commonly known simply as L-dopa or levodopa, its presence has been established in such leguminous genera as *Baptista, Lupinus, Mucuna,* and *Vicia* (Daxenbichler *et al.,* 1971), but it is not limited to this family since it accounts for 1.7% of the fresh weight of the latex secreted by *Euphorbia lathyrus* (Liss, 1961). The testa of *Vicia faba* also served as the source for the isolation of an O-β-D-glucoside of this amino acid (Nagasawa *et al.,* 1961). Later, Andrews and Pridham (1965) isolated 3-(β-D-glucopyranosyloxy)-4-hydroxy-L-phenylalanine from *Pisum sativum* seeds fed L-dopa and established that it was identical in structure to the L-dopa glucoside obtained from the testa by the Japanese workers.

A massive survey involving 724 species from 447 genera of 135 families, combined with information from other determinations, resulted in a total screening of over a thousand species. These efforts revealed that only in *Mucuna* is the L-dopa level above 0.5% of the seed dry weight (Daxenbichler *et al.,* 1971). In this genus, it was found to constitute 3.1 to 6.7% of the defatted air-dried seed meal. Bell and Janzen (1971) reported 6 to 9% free L-dopa in six species of *Mucuna* where it is the principal nonprotein amino acid of the seed. These authors also deserve credit for one of the earliest proposals for an allelochemical role for a higher plant nonprotein amino acid; their concepts are considered fully in Section G,3.

a. Biological Effects and Toxicity

L-Dopa has been implicated as a causal agent of favism, a hemolytic anemia associated with individuals possessing reduced glucose-6-phosphate dehydrogenase activity and who had also recently consumed fava beans (*V. faba*). Appreciable L-dopa is stored in this bean and an apparent relationship exists between its presence and the loss of reduced glutathione from the erythrocytes of persons deficient for this glucose-6-phosphate dehydrogenase (Kosower and Kosower, 1967). Diminished reduced glutathione appears to shorten the life span of these erythrocytes; thus, it is not unexpected that individuals with a past history of susceptibility to favism should also be deficient in glucose-6-phosphate dehydrogenase activity. On the other hand, Gaetani *et al.* (1970) have espoused a divergent viewpoint since they report normal survival rates for the red cells obtained from enzyme-deficient persons given this hydroxylated compound.

Parkinsonism is a well-known chronic neurological disorder characterized by tremors, rigidity of the limbs, and poverty of movement (hypokinesis). Widespread degenerative swelling in the basal ganglia of the brain is a common pathological manifestation of this disease (Calne and Sandler, 1970). Diminished dopamine concentration in the brain and cerebrospinal fluid is a consistent characteristic of the Parkinsonian patient (Hornykiewicz, 1966). The administration of massive quantities of L-dopa (on the order of 8 gm of the L-enantiomer per day) has had a dramatic palliative impact on the management of this syndrome. L-Dopa is decarboxylated by a low-specificity, aromatic L-amino acid decarboxylase of body tissues to produce dopamine; this provides a means for markedly increasing brain levels of dopamine since the amine cannot effectively cross the mammalian blood–brain barrier while the amino acid can. In this way, L-dopa is believed to alleviate the symptomology, particularly the hypokinesis, of Parkinson disease. Finally, intraperitoneal administration of L-dopa also produces a drastic decline in the serotonin (5-hydroxytryptamine) content of the brain (Everett and Borcherding, 1970).

b. *Metabolism*

It is reasonable to surmise that this non-protein amino acid may be formed by hydroxylation of *m*-tyrosine (3-hydroxyphenylalanine) or tyrosine (4-hydroxyphenylalanine) but the appropriate hydroxylase has not been isolated and characterized from a higher plant source. On the other hand, this enzyme is produced in rat liver which synthesizes dopa from *m*-tyrosine (Tong *et al.*, 1971). Additionally, beef adrenal medulla contains a tyrosine hydroxylase able to mediate *o*-hydroxylation of tyrosine (Shiman *et al.*, 1971). Griffith and Conn's search for a tyrosine hydroxylase in green and etiolated *Vicia faba* was not rewarded even though 26% of the applied L-[*U*-14C]tyrosine was converted to dopa by excised hypocotyls of the etiolated seedlings. Thus, while it is reasonable to surmise that dopa can be biosynthesized by simple hydroxylation of tyrosine, attempts at isolating such a tyrosine hydroxylase from selected plants proved unsuccessful (Griffith and Conn, 1973). These authors also made the point that while plant phenolases have monophenolase activity, i.e., they can catalyze *o*-hydroxylation reactions, their natural distribution is not matched by the occurrence of dopa in higher plants.

Ellis and associates have also studied L-dopa biosynthesis and believe it to be formed from tyrosine (see Remmen and Ellis, 1980). Administration of L-[*ring-U*-14C]tyrosine to young leaves of the legume *Mucuna deeringiana* resulted in the transfer of 13.7% of the total label in soluble cellular components to L-dopa. Little of the radioactive carbon was transferred to stizolobic acid. In cell suspension cultures, however, a significant conversion of the labeled tyrosine to stizolobic acid and CO_2 was reported.

Remmen and Ellis also made the interesting observation that cultured cells of this *Mucuna* species fail to accumulate L-dopa, suggesting that such manipulated cells may have lost the ability to store secondary metabolites normally occurring in high levels. A complete absence of other non-protein amino acids was also noted in cultured cells of *Lathyrus, Phaseolus,* and *Canavalia.*

Dopa has been shown to be a precursor for certain heterocyclic non-protein amino acids and this is considered in Chapter 4. An extensive elucidation of its catabolic reactions has also not been achieved. In animals, it is decarboxylated to dopamine which serves in the synthesis of such biologically important molecules as adrenaline and norepinephrine. Animals actively convert dopamine to homovanillic acid, especially in the brain.

C. BASIC COMPOUNDS

1. L-Indospicine

a. Background and Isolation.

In 1968, Hegarty and Pound announced the isolation from *Indigofera spicata* (formerly *I. endecaphylla*) of a higher plant natural product unique in having the ϵ-amidino group.

$$H_2N-C(=NH)-(CH_2)_4-CH(NH_2)COOH$$

L-Indospicine

Indospicine's isolation and eventual purification were aided by an imaginative bioassay system predicated on fat accumulation in the liver of a mouse receiving a single subcutaneous injection of the seed extract. Indospicine was isolated eventually as the sparsely soluble mono-flavianate, converted to the monohydrochloride, and crystallized from aqueous ethanol as white, needlelike crystals of the monohydrate. Culvenor *et al.* (1969) published a procedure for its chemical synthesis but it is overly cumbersome and isolation from *Indigofera* seeds remains the only practical means for obtaining this compound. A subsequent paper provided a much more detailed description of its isolation and structural elucidation and gave its concentration in mature leaves as 0.04 to 0.15% (fresh weight basis) and 0.5 to 2.0% for the seed (dry weight basis) (Hegarty and Pound, 1970).

A survey involving 17 species of *Indigofera* failed to disclose a companion source for this natural product (Miller and Smith, 1973), but its known occurrence has been extended recently to other indigoferous plants (Charlwood and Bell, 1977).

b. Biological Properties

The desirable qualities of *I. spicata* such as high crude protein content, palatability, and especially its nitrogen-fixing capability have stimulated considerable interest in its development as a forage crop but its field use has been hindered severely by its considerable abortifacient and hepatotoxic properties. A demanding examination of a thousand *Indigofera*-exposed fetuses led Pearn (1967a) to conclude that consumption of this legume causes a severe embryopathic syndrome which he characterized as consisting of a triad of recognizable lesions including clefting of the secondary palate, generalized somatic dwarfism, and varying degrees of acute hepatic toxicity. While the somatic dwarfs possess only half of the normal fetal weight, the expected proportionality prevails (Fig. 23). This adherence to anticipated body form, in all respects except for the clefting of the palate, asserts to the high site specificity of this teratogenic natural product. Indospicine's site specificity has been reaffirmed by Pearn in a strong companion paper (1967b). This is an uncommon finding for most teratogens induce a broad spectrum of embryopathic responses with little, if any, site specificity. Cleft palate formation can be induced by a single orally administered dose consisting of 1.0 ml of toxic extract per 100 gm body weight (1 ml of toxic extract contained the material obtained by extracting 10 gm of *Indigofera* seeds).

By 1970, indospicine had been isolated and synthesized and its availability allowed Pearn and Hegarty to repeat the earlier work with the pure compound. For this study, 5 rats received 2 mg per gm body weight (stomach tube) on day 13 of the gestation period. Two of the females showed intrauterine fetal death by day 21. The other three produced 16 fetuses (13 of which had cleft palates). No cleft palating was noted with the control animals which also averaged over 10 fetuses per female.

Among the nonprotein amino acids, indospicine shares with mimosine the property of being hepatotoxic. It produces liver damage in sheep, rabbits, and cows (Nordfledt *et al.*, 1952). Similarly, Sprague–Dawley rats provided 2 gm per kg body weight (stomach tube) show noticeably enlarged livers. Liver DNA, RNA, and protein contents increase marginally but the nearly 75% enhancement in organ biomass results from water retention (Christie *et al.*, 1969). Detailed histological

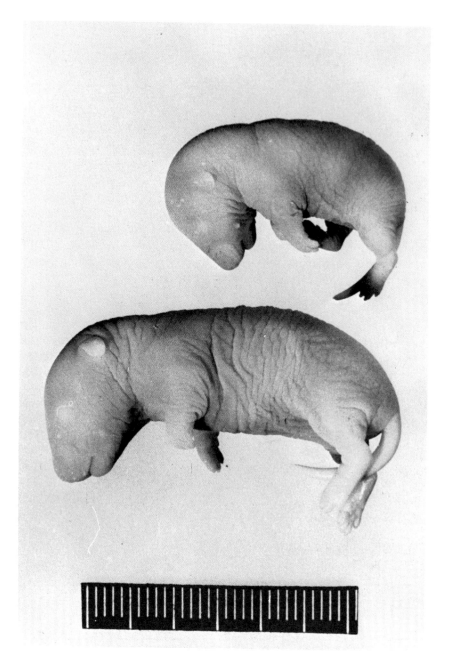

Fig. 23. Indospicine-mediated somatic dwarfism. The treated animal (top) not only has a cleft palate but also generalized somatic dwarfism. Both animals are full-term rat fetuses (scale in centimeters). Reprinted with permission from Pearn (1967a). Copyright by H. K. Lewis & Co., Ltd. Photo provided by J. H. Pearn.

studies of rats, maintained on an artificial diet supplemented with 85 to 600 mg of this basic compound per kg body weight, disclose enlarged fatty livers that develop characteristic lesions and cellular necrosis prior to the onset of cirrhosis (Christie *et al.*, 1975). In contrast, 1.6 gm canavanine per kg body weight does not affect the liver (Hegarty and Pound, 1970).

When rabbits were offered 200 gm of fresh *I. spicata* leaves as a supplement to their daily pellet ration, they responded with diminished food intake, accompanying loss in body weight, and an average survival period of only 3 weeks. Autopsies of such treated animals disclosed gross degenerative swelling and cellular necrosis of the liver. Cirrhosis was a feature common to all treated rabbits (Hutton *et al.*, 1958a). Mice consuming indospicine experience comparable liver damage; while sheep produced corneal cells of excessive opaqueness that in severe cases progressed to corneal ulcerations (Hutton *et al.*, 1958b).

c. *Toxicity*

Indospicine curtails [^{14}C]leucine incorporation into the serum and liver proteins of the rat (Christie *et al.*, 1969). To explain this observation, Madsen *et al.* (1970) examined indospicine–arginine interaction in protein production and noted that while indospicine strongly depressed aminoacylation of arginine it did not affect the charging of leucine (Table 6). That indospicine could function as an arginine antimetabolite was further demonstrated by its role as a competitive inhibitor of arginase. The K_i value for indospicine of 1.49 or 1.88 mM was quite close to arginine's K_m (1.14 mM) (Madsen and Hegarty, 1970).

It was proposed that indospicine's ability to attenuate directly arginine incorporation ultimately affected overall protein production; the latter evinced by reduced [^{14}C]leucine *incorporation*. Indospicine-mediated reduction in protein synthesis was taken as a secondary effect due to its inability to impair leucine *activation* (Madsen *et al.*, 1970). Further examination established that this arginine analogue inhibited [^{3}H]thymidine incorporation by cultured human lymphocytes; thymidine incorporation values were equated experimentally with the level of DNA synthesis. Arginine assuaged indospicine's deleterious effects; for example, a 50:1 ratio of indospicine to arginine caused a 99% reduction of thymidine incorporation while the attenuation was only 32% when the ratio fell to 4:1. A Dixon reciprocal plot provided information suggesting both competitive and reversible inhibition. Indospicine-related reduction of the thymidine incorporation rate was taken to be another ramifying effect of its antagonism of arginine metabolism and subsequent curtailment of protein synthesis (Christie *et al.*, 1971).

Table 6 Effect of Indospicine *in Vitro* on the Charging of tRNA with L-[U-^{14}C]Arginine or DL-[1-^{14}C]Leucine[a]

Sample	Indospicine (mM)	Activity (cpm/sample)	Percentage of control value
L-[U-^{14}C]Arginine	None	780	100
	0.1	780	100
	1.0	655	84
	2.5	380	49
	5.0	218	28
	10.0	0	0
DL-[1-^{14}C]Leucine	None	2640	100
	10.0	2680	101

[a] Reproduced with permission from the research of Madsen *et al.* (1970). Copyright by Pergamon Press, Ltd.

Unfortunately, data are not available on the biosynthesis of the novel amidino group nor is it known if or how this metabolite provides nitrogen for the intermediary metabolism of the producer plant.

2. L-Canavanine

a. Background and Isolation

As part of a series of studies spanning a period from 1929 through 1937, Kitagawa and associates isolated and chemically characterized a novel basic amino acid of plants. They gave it the trivial name canavanine after its isolation from *Canavalia ensiformis* (jack bean) (Kitagawa and Tomiyama, 1929).

$$H_2N—C(\!\!=\!\!NH)—NH—O—CH_2—CH_2—CH(NH_2)COOH$$

L-Canavanine

Several purely chemical preparations as well as isolation from biological sources have been developed (Gulland and Morris, 1935; Damodaran and Narayanan, 1939), but ion-exchange chromatography of a seed extract has proved markedly superior for large-scale production of the L-enantiomer (Hunt and Thompson, 1971; Rosenthal, 1973, 1977b).

Appreciable interest exists in canavanine's natural occurrence and distributional pattern, largely derived from its considerable application to questions of chemotaxonomy. In this regard it has proved of real value in understanding and delineating phylogenetic relationships in the Papiloinoideae (Fabaceae) (Lackey, 1977; Bell *et al.*, 1978). This large

Table 7 General Biological Effects of L-Canavanine[a]

Organism	Biological effects
Encephalomyelitis virus	Inhibited growth
Lee influenza virus	Inhibited growth
Streptococcus faecalis var. *liquefaciens*	Stimulated growth in the presence of arginine
Streptococcus bovis	Stimulated growth under low carbon dioxide and arginine concentrations
Staphylococcus aureus	Decreased enzyme activity
Anabaena flos-aquae	Reduced nitrogenase synthesis
Chlamydomonas reinhardi	Inhibited growth and gametogenesis
Saccharomyces, Torulopsis, and other yeastlike fungi	Severe growth reduction
Saccharomyces cerevisiae	Reduced respiratory activity (mitochondrial cytochromes *a* and *b* synthesis significantly reduced)
Saccharomyces cerevisiae and *S. carlsbergenesis*	Stimulated amino acid-dependent proton uptake
Candida albicans	Decreased survival of ultraviolet-treated cells
Puccinia recondita	Reduced rust development
Puccinia coronata	Prevented rust pustule formation
Peronospora tabacina	Complete inhibition of sporulation
Aspergillus flavus	Disrupted differentiation
Coprinus lagopus	Inhibited growth
Ustilago maydis (arginine auxotroph)	Increased frequency of forward mutation and revertants
Oats	Inhibited indoleacetic acid-dependent coleoptile elongation
Corn (cultured embryos)	Inhibited growth
Vicieae (legumes)	Reduced pollen germination and pollen tube growth
Carrot (explants)	Inhibited growth
Tobacco (cultured cells)	Inhibited growth
Phaseolus vulgaris (bean)	Reduced arginine utilization and initiation of mitosis; other mitotic events also affected
Cattleya (orchid)	Reduced seed germination and growth
Bean, corn, soya bean	Inhibited root growth
Tetrahymena pyriformis	Inhibited RNA synthesis
Mouse embryonic cells	Reduced cellular proliferation
Mouse L-929 cells	Destroyed arginine-starved interferon-treated cells
Sheep erythrocytes	Reduced complement-dependent lysis
Chick embryonic fibroblasts	Decreased survival
Rat liver and cerebral cortex mitochondria	Decreased protein and glycoprotein synthesis
Walker carcinosarcoma 256 cells	Competitively inhibited arginine uptake
Assorted cultured human cells	Prevented cellular proliferation
Humans	Inhibited placental alkaline phosphatase activity

[a] See original for references. After Rosenthal (1977a). Copyright by the *Quarterly Review of Biology*.

group of leguminous plants continues to be the sole confirmed sources of this secondary metabolite in spite of occasional and reoccurring assertions to the contrary (see Rosenthal, 1977a,1978c; Bell *et al.*, 1978).

The recent report of a massive examination conducted by Bell *et al.* (1978) has extended the known distribution of this secondary metabolite to 500 species representing 240 genera. Their efforts established 70 genera as new sources of this arginine analog in the Papiloinoideae and represents the most definitive chemotaxonomic study of canavanine conducted to date.

My experience with a hundred or so of these species indicate that most seeds contain about 2 to 4% of this metabolite. Many have much smaller amounts, on the order of 0.1 to 0.2%, but others are veritable canavanine store houses; concentrations of 10 to 13% of the seed dry weight are no longer an unusual finding. In certain legumes, the total proportion of seed nitrogen allocated to canavanine can easily exceed 90% of the total soluble amino acid nitrogen (see Table 15).

Analyses of leguminous seeds by numerous workers, commencing with the 1960 report of Bell, have revealed that seed canavanine declines precipitously during seed germination. All detectable canavanine is exhausted before the cotyledons wrinkle and abscise. Canavanine's depletion as a function of growth, abundance, high nitrogen content, and significant nitrogen to carbon ratio have resulted in a consensus that one of canavanine's prime functions is that of a nitrogen-storing metabolite in the seed. The principal higher plant catabolic pathway for mobilization of the nitrogen stored in canavanine involves sequential hydrolytic cleavages mediated by arginase and urease. By these reactions, canavanine is converted to canaline and urea prior to degrading the latter to carbon dioxide and ammonia. In jack bean, chromatographic evidence has been obtained for a subsequent diversion of nitrogen into the amide group of asparagine, which accumulates in the aging cotyledons prior to its presumed translocation to the vegetative tissues of the growing plant (Rosenthal, 1970).

b. *Biological Properties and Toxicity*

Canavanine's adverse biological effects have been demonstrated in an unusually wide variety of organisms (Table 7) but these studies are largely of a descriptive nature and as such have provided limited insight into its basic toxicology. The potent antimetabolic properties of canavanine result primarily from its ability to function as a highly effective antagonist of arginine metabolism due to its structural similarity to this protein amino acid.

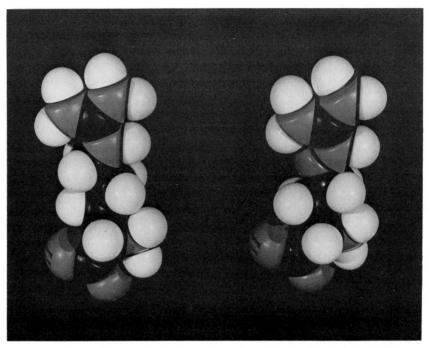

CPK space-filling model of L-canavanine (right) and L-arginine (left).

Canavanine represses the enzymes mediating arginine biosynthesis in *E. coli* (Maas, 1960), serves as a substrate in virtually all enzyme-catalyzed reactions that preferentially use arginine, and competitively inhibits the arginine uptake and transport system of organisms in which it has been studied (see Rosenthal, 1977a).

Canavanine can be aminoacylated by arginyl-tRNA synthetase (Allende and Allende, 1964; Mitra and Mehler, 1967); evidence of canavanine incorporation into protein has been obtained in adeno-virus (Neurath *et al.*, 1970), *E. coli* (Schachtele and Rogers, 1965), *Chlamydomonas reinhardii* (McMahon and Langstroth, 1972), Walker car-cinosarcoma cells (Kruse *et al.*, 1959), and in the tobacco hornworm, *Manduca sexta* (Dahlman and Rosenthal, 1975). When canavanine re-places arginine, a less basic residue is positioned in the polypeptide chain since canavanine is isoelectric at 8.2 as compared to 10.8 for ar-ginine (Greenstein and Winitz, 1961). The resulting decrease in basicity can alter residue interaction and ultimately disrupt tertiary and/or quaternary structure. In support of this concept, canavanine blocks sub-unit dimerization, a prerequisite for catalytically active alkaline phosphatase (Attias *et al.*, 1969). Regrettably, definitive information on the exact nature of canavanine-mediated alteration of catalytic or regulatory enzyme properties has not been obtained.

All arginine-containing proteins are subject potentially to canavanine replacement and the great diversity of canavanine's effects in so many distinctive organisms suggests that widespread replacement in many types of proteins undoubtedly occurs. The proteins of a canavanine-sensitive organism provided this toxic natural product have never been found to be canavanine-free; the converse is equally true since canavanine-producing legumes do not synthesize detectable canavanyl proteins (unpublished observations; see also Fowden and Frankton, 1968). While this proves nothing, it is consistent with the mounting evidence implicating canavanine toxicosis with aberrant protein production.

In prokaryotes, canavanyl proteins form a conjugate with the bacterial genome which prevents further cycles of DNA replication (Schachtele *et al.*, 1970; Neurath *et al.*, 1970). Studies of axenically grown soya bean roots (Weaks and Hunt, 1973) indicate that overall DNA metabolism, as assessed by [^{3}H]thymidine incorporation into trichloroacetic acid-insoluble material, is attenuated by this arginine analog.

Study of Semliki forest virus cultured on BHK 21 cells (Ranki and Kääriäinen, 1969, 1970) have provided experimental evidence for direct canavanine-mediated curtailment of RNA polymerase synthesis. If canavanine can significantly diminish mRNA production, then once existing mRNA molecules are degraded, the translational activity of the cell must be markedly hampered. Proteins required for genomic replication and expression would all be affected adversely; this type of domino theory may be applicable in certain instances of canavanine toxicity.

c. Metabolism

The metabolic reactions culminating in canavanine's biosynthesis have not been fully established but certain aspects have been clarified. Each of the components of the classical Krebs–Henseleit ornithine–urea cycle has a nonprotein amino acid counterpart in which the terminal methylene group is replaced by oxygen (Fig. 24). Jack bean produces all of the ornithine–urea cycle enzymes and they are able to foster the reactions depicted in Fig. 24 (see Rosenthal, 1972).

Injection of [^{14}C]carbamoyl phosphate into the fleshy cotyledons of 8-day-old jack bean plants resulted in appreciable labeling of both the arginine and canavanine contained within the trichloroacetic acid-treated plant extract. Analysis of these two basic amino acids 6, 12, and 24 hr after injection revealed that 27.1, 21.2, and 14.2%, respectively, of

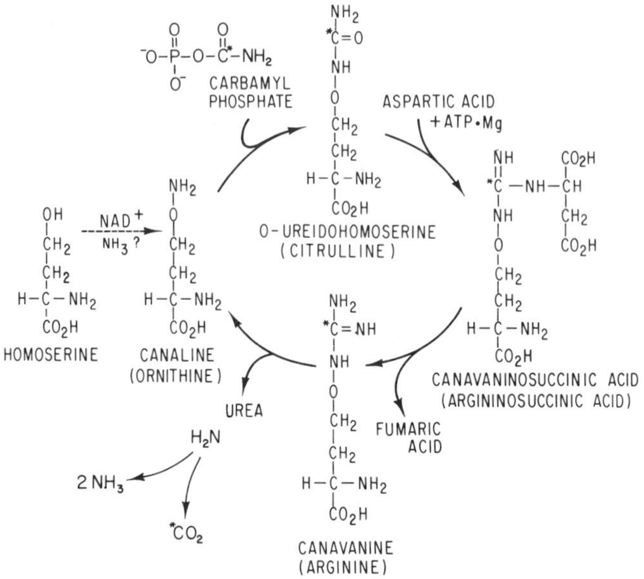

Fig. 24. Reactions of canavanine metabolism in jack bean, *Canavalia ensiformis*. This reaction sequence involves a series of nonprotein amino acids bearing structural analogy to the constituents of the Krebs–Henseleit ornithine–urea cycle (the latter are indicated in parentheses). The arrows denote the predominate reaction flow; only the conversion of canavanine to canaline and urea is irreversible. The asterisk next to various carbon atoms highlights the movement of radioactive carbon from carbamoyl phosphate to its release as $^{14}CO_2$. The role of homoserine or one of its derivatives in providing carbon skeleton to this reaction sequence has not been firmly established.

the soluble radioactivity of the plant extract resided in canavanine and arginine (Table 8).

Concurrent enzymatic hydrolyses of the isolated radioactive canavanine and arginine with arginase and urease revealed that virtually all the radioactivity was contained in the terminal guanidinooxy and guanidino group, respectively. This is exactly the isotope transfer pattern necessitated by the carbamoyl phosphate-dependent formation of canavanine postulated in Fig. 24. When 25 μmoles of canaline were provided with labeled carbamoyl phosphate, the radioactivity in the newly formed canavanine isolated after 12 hr was increased 48% as compared to [^{14}C]carbamoyl phosphate only. This enhancement in labeled carbon transfer to canavanine resulted from increased carbamylation of canaline which is in direct competition with ornithine for the formation of arginine as well as aspartic acid which diverts carbamoyl

Table 8 Formation of L-[*Guanidinooxy*-¹⁴C]Canavanine and L-[*Guanidino*-¹⁴C]Arginine from [¹⁴C]Carbamoyl Phosphate by Intact Jack Bean[a]

Time (hr)	Plant extract (dpm × 10⁻⁶)	¹⁴C incorporation (% total soluble radioactivity)		Arginine/canavanine
		Arginine	Canavanine	
6	4.37	19.8	7.3	2.71
12	3.56	15.1	6.1	2.48
24	1.98	9.8	4.4	2.23

[a] The values are expressed on a per plant basis. Each of four plants received 5 μCi of [¹⁴C]carbamoyl phosphate. The radioactivity of the plant extract is based on soluble radioactivity after trichloroacetic acid treatment and extraction. Taken from Rosenthal (1982). Copyright by the American Society of Plant Physiologists.

phosphate into the reactions of pyrimidine metabolism. Once again, essentially all the radioactive carbon of canavanine produced from [¹⁴C]carbamoyl phosphate occurred exclusively in the guanidinooxy moiety of this arginine analog. This finding is in complete agreement with the contention that canaline, in conjunction with carbamoyl phosphate, supports the reactions of Fig. 24 that culminate in canavanine biosynthesis.

Attempts were also made to produce labeled canavanine from other compounds not associated directly with canavanine production. No significant labeling of canavanine occurred with canaline and L-[*guanidino*- ¹⁴C]-arginine which suggests that this transamidination reaction may not be of importance in canavanine formation. Radioactive glutamic acid, glutamine, aspartic acid, asparagine, urea, acetate, and bicarbonate all failed to transfer significant radioactive carbon to canavanine. Feeding experiments with L-[*U*-¹⁴C]homoserine led to discernible labeling of the lysine pool of jack bean, perhaps via the conversion of homoserine to aspartic semialdehyde which is then transformed to lysine via 2,3-dihydropicolinic acid and *meso*-2,6-diaminopimelic acid. In a typical study that spanned 12 hr, 2.9% of the administered carbon occurred in lysine but only 0.7% in canavanine. It is possible that the administered homoserine did not equilibrate significantly with compartmentalized homoserine that supports the reactions of canavanine biosynthesis from canaline (see Rosenthal, 1982).

In summary, it is evident that jack bean mediates a reaction sequence from canaline that culminates in canavanine biosynthesis and these reactions are best viewed as the synthetic phase of canavanine metabolism. The combined hydrolytic actions of arginase and urease mobilize the nitrogen sequestered in the guanidinooxy moiety of canavanine by the sequential formation of canaline and urea and CO_2 and ammonia (Rosenthal, 1970). In fact, a correlation exists between seed canavanine content and the urease activity found in these tissues (Rosenthal, 1974). This hydrolytic reaction sequence is properly viewed as the degradative phase of canavanine metabolism. Thus, the reactions of canavanine metabolism are best viewed as having a synthetic phase in which assimilated nitrogen funnels from carbamoyl phosphate and aspartic acid into canavanine with canaline functioning as the nitrogen recipient. During the degradative phase, canavanine is converted to canaline and urea. Urea is broken subsequently to provide reduced nitrogen as ammonia for a myriad of higher plant biosynthetic reactions. Nothing is known definitely of canaline catabolism but it has been proported to serve as a source of homoserine (Töpfer et al., 1970; Johnstone, 1956). The reactions of the synthetic phase would be expected to dominate during canavanine synthesis and translocation concomitant with flower anthesis and seed development. The catabolic phase would be expected to predominate during seed germination and development when stored nitrogen is required for the growth of the plant. This point is being tested experimentally at this time.

d. Canavanine Detoxification

Studies of the neotropical bruchid beetle, *Caryedes brasiliensis*, have provided substantive biochemical insights into insect detoxification of a normally poisonous higher plant constituent. *Dioclea megacarpa*, a leguminous species of the deciduous forests of Costa Rica, produces a seed storing up to 13% of its dry matter as canavanine (Rosenthal et al., 1976b). The seeds of this plant represent the principal foodstuff for the developing larvae and their canavanine-containing tissues are ravaged by the time the adult beetles emerge from their larval food source (Fig. 25).*

Consumption of appreciable canavanine-laden tissue usually represents a severe biological hazard since the arginyl-tRNA synthetase of non-producing species activate canavanine. This ultimately results in the synthesis of anomalous canavanyl proteins. A major biochemical adaptation of this insect to the appreciable canavanine in its foodstuffs is a strict avoidance of canavanyl protein production; this is achieved by the production of an arginyl-tRNA synthetase able to discriminate between canavanine and arginine. In this way, canavanine is not activated nor covalently linked to the growing polypeptide chain (Rosenthal et al.,

Dioclea wilsonii is also consumed by this seed predator.

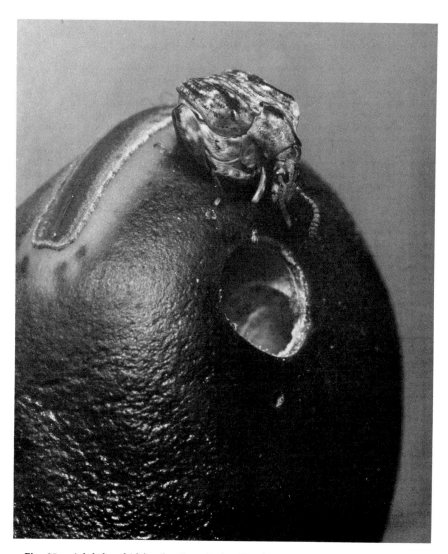

Fig. 25. Adult bruchid beetle, *Caryedes brasiliensis*, recently emerged from a *Dioclea* seed. The adult's exit hole and the small entrance orifice of the first stadium larva are shown. Developing bruchid beetle larvae feed on the *Dioclea* seed in spite of the presence of appreciable canavanine in the seed. Reprinted with permission from Rosenthal *et al.* (1977a). Copyright by Stony Brook Foundation, Inc. Photo provided by D. H. Janzen.

1976b). This discriminatory ability represents a major biochemical adaptation of the bruchid beetle to this toxic food constituent.

Analysis of the free amino acids of the *Dioclea* seed reveals that canavanine constitutes 96% of the total ninhydrin-positive compounds (see Table 15, p. 153). Bruchid beetle larvae generally exhibit limited proteolytic activity (Applebaum, 1964) which limits seed protein as an abundant source of dietary amino acids. These factors formed the basis for the belief that rather than simply excreting, detoxifying, or otherwise avoiding this poisonous compound, this seed predator diverts canavanine into a dietary source of nitrogen. Bruchid beetles share with other insects demonstrative arginase activity which provides ornithine for other metabolic reactions such as polyamine synthesis and proline formation. This enzyme also provides the means for hydrolysis of canavanine to canaline and urea. Urea's nitrogen cannot be assimilated

Table 9   Urease Content of Various Terminal-Stadium Insect Larvae[a]

Organism	Urease content (μU/mg)
Diptera	
Drosophila melanogaster	15
Musca domestica	95
Hymenoptera	
Caliroa sp.	ND
Lepidoptera	
Manduca sexta	ND
Hyalophora cecropia	ND
Hyphantria cunea	ND
Heliothis virescens	ND
Pseudoaletia unipuncta	110
Galleria mellonella	ND
Ephestia kühniella	70
Coleoptera	
Tribolium castaneum	ND
Anthonomus grandis	ND
Hypera postica	480
Leptinotarsa decemlineata	1,405
Callosobruchus maculatus	1,245
Caryedes brasiliensis	38,570

[a] Urease content is expressed as microunits per milligram of soluble protein; ND denotes organisms lacking detectable urease activity. See original for experimental details. Reproduced from the work of Rosenthal *et al.* (1977). Copyright by the American Association for the Advancement of Science.

directly but it can be utilized as ammonia. Urease, responsible for this conversion, is rarely reported in insects and accounts of its occurrence in these organisms have been discounted generally (Cochran, 1975). Nevertheless, this insect not only produces urease but nearly 30 times the amount found in other tested insect species (Rosenthal *et al.*, 1977) (Table 9). Thus, this insect surmounts the obstacles created by the disproportional sequestration of seed nitrogen into canavanine and the limited availability of seed storage protein by the sequential conversion to canaline and then urea and finally ammonia.

Caryedes brasiliensis has appreciable ammonium-dependent glutamine synthetase activity and contains a high level of glutamine relative to the other free amino acids. This reaction pathway represents a means of utilizing ammonia released from urea as well as sequestering this toxic cation in an innocuous and utilizable form. These biochemical abilities represent a second series of biochemical adaptations that results in the conversion of a highly toxic plant product such as canavanine into a utilizable nitrogenous resource. Release of canavanine's stored nitrogen by urea formation also produces a stoichiometric release of canaline, a highly toxic natural product in its own right (see Section C,3). This particular predicament is circumvented biochemically by the reductive deamination of canaline to homoserine and ammonia (Rosenthal *et al.*, 1978).

$$H_2N—C(=NH)—NH—O—CH_2—CH_2—CH(NH_2)CO_2H \longrightarrow \qquad (i)$$
L-Canavanine

$$H_2N—O—CH_2—CH_2—CH(NH_2)CO_2H \; + \; H_2N—C(=O)—NH_2$$
L-Canaline $\qquad\qquad\qquad\qquad\qquad\qquad$ Urea

$$H_2N—C(=O)—NH_2 \longrightarrow \quad CO_2 + 2NH_3 \qquad\qquad (ii)$$
Urea

$$H_2N—O—CH_2—CH_2—CH(NH_2)CO_2H \; + \; NADH \; + \; H^+ \longrightarrow \quad (iii)$$
L-Canaline

$$HO—CH_2—CH_2—CH(NH_2)CO_2H \; + \; NH_3 \; + \; NAD^+$$
L-Homoserine

$$HOOC—CH_2—CH_2—CH(NH_2)COOH \; + \; ATP·Mg \; + \; NH_3 \longrightarrow \quad (iv)$$
L-Glutamic acid

$$H_2N—C(=O)—CH_2—CH_2—CH(NH_2)COOH \; + \; ADP·Mg \; + \; P_1$$
L-Glutamine

By conducting integrated reactions (i) through (iv), *C. brasiliensis* detoxifies canavanine and canaline, provides amide nitrogen for metabolic reactions, and conserves the basic carbon skeleton as homoserine. These reactions are also consistent with the appreciable free homoserine and glutamine found in a trichloroacetic acid-precipitated extract of bruchid beetle larvae (unpublished observation).

The results of these studies formed the basis of a hypothesis presented by Rosenthal *et al.* (1978) that this bruchid beetle has adapted to the toxic components of *Dioclea* by functioning biochemically much as a canavanine-synthesizing plant. This example of convergent evolution is supported by several lines of evidence. First, the production by the insect of an arginyl-tRNA synthetase able to distinguish between arginine and canavanine. Higher plants avoid nonprotein amino acid autotoxicity by having aminoacyl-tRNA synthetases able to discern between an amino acid that is a protein constituent and one that is not (Fowden and Lea, 1979). Second, the observation that urea is produced from canavanine followed by urease-mediated ammonia formation. This biochemical sequence is also the principal catabolic pathway for mobilizing canavanine's nitrogen by legumes (Rosenthal, 1970). Another parallel is revealed by the recent elucidation of canaline detoxification by conversion to homoserine. Canavanine synthesis in the jack bean plant proceeds by a reaction sequence involving canaline (Fig. 24). Canaline formation may involve homoserine as a precursor. Thus, while the purpose for the interconversion of canaline and homoserine differ in the bruchid beetle from that of the canavanine-synthesizing legume, these nonprotein amino acids share a vital role in the basic reactions characteristic of canavanine metabolism. Finally, there is the sequestration of the nitrogen in ammonia in the form of an amide nitrogen-containing amino acid. This represents an established plant means for storage and utilization of ammonia.

Recent examination of the frass of *C. brasiliensis* revealed that uric acid, the principal nitrogen excretory metabolite of insects, accounts for only 11% of the nitrogen (Rosenthal and Janzen, 1981). Ammonia and urea constituted 42% and 47%, respectively of the excreted nitrogen of the frass. Thus, this seed predator also emerges as being unusual in its strong ureotelic and ammoniotelic elimination of nitrogen.

Simultaneous development of the biochemical capacities described above represent an improbable event but one rendered more rational when viewed as a series of adaptive occurrences. Perhaps a progenitor of *C. brasiliensis* was able to avoid canavanyl protein production as its extant offspring does but had not, as yet, adapted to canavanine or canaline. While the fitness of the early invaders of this food resource undoubtedly fell significantly, they survived! In time, the remnant population acquired the ability to manipulate canavanine. Canaline also came to be degraded eventually; it represented a means of detoxifying this poisonous natural product while enhancing the efficacy of carbon and nitrogen utilization from stored canavanine. Sequential acquisition of these biochemical capabilities continued to increase the overall biological fitness of this seed predator as it became better adjusted and more committed to its new food resource.

It is possible that the several accomodations inherent in the total adaptive process occurred over a very considerable time period. That is, early *D. megacarpa* may have stored little canavanine but seed predator pressure instigated an ever higher commitment of the seed nitrogen resource to canavanine production. The ever increasing concentration of canavanine gradually selected for those individuals able to accomodate the increasing level of this seed nonprotein amino acid.

By ovipositing mostly on *D. megacarpa*, the female limits drastically the diversity, availability, and extent of its larval food supply. This immediately raises the question: Why hasn't this predator moved to other canavanine-containing legumes of its habitat? This is particularly relevant since this herbivore has already invested in development of the metabolic mechanisms to fully exploit canavanine. Perhaps other canavanine-containing legumes of this habitat store toxic compounds to which *C. brasiliensis* has not adapted or it may be unresponsive to the physical and/or chemical cues critical to locating and recognizing potential canavanine-producing hosts. Being the only known predator of this legume, *C. brasiliensis* most certainly benefits from unencumbered access to its food source; the usual competitive struggle for food resources need not concern this beetle. In addition, it only needs a limited detoxification capability instead of a wide array of metabolically costly modes of response. While these factors represent cost-benefit items favoring the monophagous feeding habit of this seed predator, they severely minimize its ability to radiate into new food territories.

The interaction of *C. brasiliensis* with this leguminous species is an excellent example of biological compromise in which certain benefits from lack of competition or the need to maintain diversified detoxification strategies must be weighed against precarious dependency upon a single food resource. It would be most interesting to observe the effects of a long term loss of *D. megacarpa* from the habitat and to follow the biochemical response of surviving members to the attrition of their food and ovipositing material.

3. L-Canaline

a. Background and Isolation

Initial awareness of canavanine's existence resulted from the fact that in Kitagawa's laboratory a jack bean seed extract provided the urease employed for urea determinations. The resulting values were always too high relative to other methods of analysis since extraneous urea was generated by enzymatic cleavage of canavanine. Their search for the cause of the errant urea values led to canaline's discovery since it is also a stoichiometric reaction product of canavanine degradation (Kitagawa and Yamada, 1932). Canaline, the only known naturally occurring nonprotein amino acid having a free aminooxy group, bears structural analogy to ornithine.

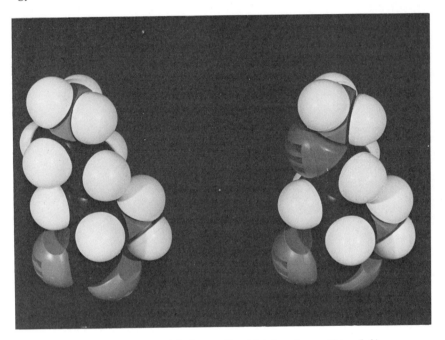

CPK space-filling model of L-canaline (right) and L-ornithine (left).

A substituted hydroxylamine, canaline can decompose completely in organic solvent systems used to isolate it by partition and ion-exchange chromatography (see Rosenthal, 1978a). This property has minimized severely its study from natural sources. Canaline has limited resistance to autoclaving but its overall stability has not been systematically investigated.

Canaline has been isolated from leguminous materials (Miersch, 1967) and all the 500 species known to produce canavanine (and presumably arginase as well) represent potential natural sources for this compound.

b. Biological Properties

Few investigations of the biological properties of canaline have been made; for example, whole animal studies are limited solely to the tobacco hornworm, *Manduca sexta*. Canaline exhibits potent insecticidal properties when administered to *M. sexta* adults. It also produces depauperate larvae, enhances pupal and adult malformations, attenuates survival of all developmental stages, and significantly curtails ovarial mass production by gravid females (Rosenthal and Dahlman, 1975). Certain neuropharmacological properties have been disclosed by studies of the adult moth (Kammer *et al.*, 1978) where it induces almost continuous motor activity. The treated moth flies normally at first but rapidly becomes disoriented; muscle activity is less patterned even though wing movement continues. Neither axonal conduction of action potentials nor the activity of mechanoreceptors is affected, but the postsynaptic potential of flight muscle fibers is prolonged. This particular effect, commencing 2 to 10 min after the application of canaline, is of a transitory nature and after 20 to 40 min the response of the muscle fibers to electrical stimulation of the motor nerve is normal.

The structural similarity of canaline to γ-aminobutyric acid (GABA) is apparent immediately.

$$
\begin{array}{cc}
\mathrm{NH_2} & \\
| & \\
\mathrm{O} & \mathrm{NH_2} \\
| & | \\
\mathrm{CH_2} & \mathrm{CH_2} \\
| & | \\
\mathrm{CH_2} & \mathrm{CH_2} \\
| & | \\
\mathrm{H-C-NH_2} & \mathrm{H-C-H} \\
| & | \\
\mathrm{COOH} & \mathrm{COOH}
\end{array}
$$

L-Canaline   L-γ-Amino-
butyric acid

It is reasonable to conjecture that canaline antagonizes GABA-mediated biological actions in the central nervous system. However, when picrotoxin, a known inhibitor of GABA, is injected into the adult moth, it does not mimic the biological effects induced by canaline. Activation of the flight pattern generator, a continuous motor output, and prolongation of postsynaptic potential suggest that canaline affects the central nervous system. While canaline's exact mode of action remains unknown, it does disrupt central nervous system function and, as such, enlarges the list of known neurotoxic plant nonprotein amino acids.

The potent growth-reducing properties of canaline have been documented in a series of experiments employing the aquatic microphyte, *Lemna minor* (Rosenthal *et al.*, 1975). Addition of only 10^{-6} *M* canaline to the sterile, chemically defined medium utilized for growing this higher plant produces nearly a 20% attrition in growth (i.e., frond production). When the canaline level is increased to 10^{-5} *M*, the fronds serving as inoculum fail to proliferate; higher levels bleach the fronds white. If ornithine, citrulline, or arginine are provided at a 10-fold molar excess relative to canaline, the fronds are protected fully. At an equimolar level, however, these amino acids cannot diminish canaline's toxicity (Rosenthal *et al.*, 1975). A particularly interesting finding of the *Lemna* growth studies was the discovery of the potent additive and synergistic growth inhibiting properties of the amino acids of canavanine synthesis. Canavanine and canaline suppress *Lemna* growth additively while ureidohomoserine exerts a profound synergistic effect on both their individual and synergistic growth retarding capabilities (Fig. 26). Even the intrinsic toxicity of canavaninosuccinic acid, by far the least toxic of these compounds, is amplified by the presence of ureidohomoserine. As a result, these compounds acquire a remarkable degree of *collective* potency (Table 10) (Rosenthal *et al.*, 1976a).

Canaline is also a very powerful inhibitor of pyridoxal phosphate-containing enzymes. Rat glutamate-oxaloacetate and glutamate-pyruvate transaminases are inhibited almost completely by treatment with only 10 mM canaline (Kekomäki *et al.*, 1969). Ornithine-ketoacid transaminase-mediated production of glutamate is curtailed 86 and 100% by 5 μM and 50 μM canaline, respectively (Katunuma *et al.*, 1965). The former group of workers also examined canaline's effect on urea production since the ornithine antagonist, 2,4-diaminobutyric acid drastically reduces urea production; no such effect is noted with canaline (Kekomäki *et al.*, 1969). Ornithine interaction with canaline has been examined with ornithine carbamoyltransferase of human liver; neither compound significantly inhibits this enzyme when the other amino acid functions as the carbamoyl group recipient (Natelson *et al.*, 1977). While canaline strongly attenuates the activity of all seven pyridoxal phosphate-containing enzymes examined by Rahiala *et al.* (1971), it does not affect three ornithine-utilizing enzymes lacking a B$_6$ cofactor. In the jack bean plant, ornithine carbamoyltransferase vigorously forms O-ureidohomoserine from canaline and carbamoyl phosphate, yet O'Neal (1975) reported that canaline and ornithine do not inhibit production of citrulline and ureidohomoserine, respectively. Thus, it appears that the reaction of canaline with the bound coenzyme of pyridoxal phosphate-containing enzymes is the principal basis for the antimetabolic properties of canaline.

This conclusion is also supported by an experimental finding suggesting that canaline undergoes oxime formation with pyridoxal phosphate by reacting with the aldehyde group of the coenzyme (Rahiala *et al.*, 1971; Rahiala, 1973). The suggestion of such a complex actually has its foundation in the work of DaVanzo *et al.* (1964) who noted that aminooxyacetic acid, structurally similar to canaline, formed an oxime

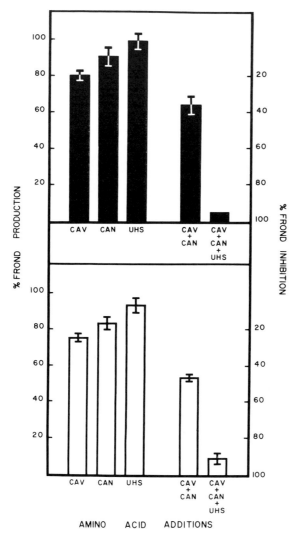

Fig. 26. The additive and synergistic properties of canavanine, canaline, and ureidohomoserine. Frond production, expressed as a percentage of the control culture growth, represents the mean of eight cultures ±SE and was determined after 6 days (open bars) and 10 days (solid bars) of growth. The amino acids were each provided at a $1\mu M$ concentration. CAV, CAN, and UHS denote canavanine, canaline, and ureidohomoserine respectively. Reprinted with permission from Rosenthal *et al.* (1976a). Copyright by the American Society of Plant Physiologists.

Table 10   The Combined Effect of Canavanine, Canaline, and Ureidohomoserine on the
Growth of *Lemna minor*[a]

Concentration (μM)	Frond production (% of control)	
	6 days	10 days
1.00	Lethal	Lethal
0.50	32.9 ± 4.4	58.0 ± 6.2
0.25	84.9 ± 2.9	92.0 ± 2.8
0.125	93.9 ± 2.6	95.3 ± 3.4

[a] Each value, expressed as a percentage of the control, represents the mean of eight cultures ±SE. The control cultures averaged 23.6 ± 0.8 fronds at 6 days and 101.1 ± 4.8 fronds at 10 days. Each amino acid was supplied at the indicated concentration. Reproduced with permission from Rosenthal *et al.* (1976a). Copyright by the American Society of Plant Physiologists.

type complex with the coenzyme. Experimental confirmation of this belief was provided by a study of the pyridoxal phosphate group of rat liver cystathionase (Beeler and Churchich, 1976). Addition of either aminooxyacetate or canaline to the enzyme eradicated the 420 nm absorption band believed to result from covalent interaction between the coenzyme and the apoenzyme. A novel peak appeared at 330 nm; this peak is created by the reaction between canaline and pyridoxal phosphate (Fig. 27A). These experimental findings resulted in the assertion that aminooxy-containing compounds disrupt the Schiff base linkage essential for complexing pyridoxal phosphate to the apoenzyme.

The canaline–pyridoxal phosphate complex has been prepared and chemically characterized (Rosenthal, 1981). Spectral studies reveal that the reaction of canaline with the free cofactor quenches the 390 nm peak of pyridoxal phosphate and produces a novel peak at around 330 nm (Fig. 27B).

When canaline is reacted with bacterial L-tyrosine decarboxylase, a known pyridoxal phosphate-containing enzyme, a concentration-dependent inactivation of the enzyme's decarboxylation ability ensues. Addition of increasing amounts of canaline to the holoenzyme not only attenuates the enzymatic activity but also produces a series of increasingly stronger spectral shifts that are in accord with enhanced formation of the canaline–pyridoxal phosphate complex (Fig. 27B).

When canaline-treated holoenzyme is dialyzed exhaustively, the absorption spectrum returns to that of the apoenzyme. This canaline-treated but dialyzed enzyme can be reactivated by treatment with 5 mM pyridoxal phosphate. This finding suggests that, as with pyridoxal phosphate itself, the canaline-pyridoxal phosphate complex is bound reversibly to the holoenzyme.

It is reasonable to propose that a biochemical basis for the marked antimetabolic properties of canaline reflects its ability to complex with the B_6-moiety of an enzyme thereby rendering the enzyme inactive. NMR data obtained from this complex indicate the involvement of a Schiff base in the formation of this reversibly bound complex (Rosenthal, 1981a).

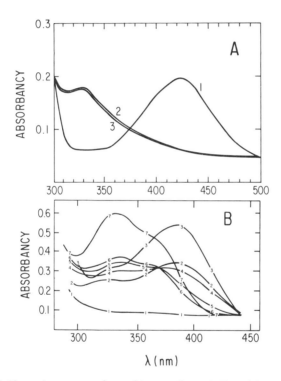

Fig. 27. (A) Absorption spectra of cystathionase. Curve 1, Cystathionase only; curve 2, aminooxyacetic acid (0.4 mM); curve 3, canaline (50 μM). The enzyme (6 μM) was reacted with these reagents at pH 7.4 for 10 min prior to spectral determinations. Reprinted with permission from Beeler and Churchich (1976). Copyright by American Society of Biological Chemists, Inc. (B) Spectral scans. The absorbance of the indicated samples is shown as a function of wavelength. All determinations were conducted at pH 5.5 in 100 mM sodium acetate buffer. Curve 1, L-Tyrosine decarboxylase apoenzyme; 2, L-tyrosine decarboxylase holoenzyme; 3, free pyridoxal phosphate; 4, holoenzyme plus 0.1 mM canaline; 5, holoenzyme plus 0.2 mM canaline; 6, holoenzyme plus 0.4 mM canaline; 7, canaline–pyridoxal phosphate complex. Reproduced from Rosenthal (1981).

4. L-Homoarginine

a. Background and Isolation

Inclusion of the higher homolog of arginine among higher plant natural products results from the nearly concurrent reports of its isolation by Bell (1962a,b) and Rao *et al.* (1963). *Lathyrus cicera* (Bell, 1962c) served subsequently as the plant source for a chemical characterization of the isolated compound.

$$H_2N—C(=NH)—NH—(CH_2)_4—CH(NH_2)COOH$$

L-Homoarginine

Elucidation of homoarginine's identity permitted Bell to assign retrospectively this nonprotein amino acid to 36 species of *Lathyrus* from data obtained in an earlier chromatographic study involving 49 species of the genus (Bell, 1962a). It is presently known from 40 species of *Lathyrus*, of which 16 also store its γ-hydroxylated derivative and lathyrine (Bell, 1964).

Rao *et al.* (1963) obtained their sample of this basic compound from *L. sativus* by purifying an extract of 3 kg of seed meal utilizing Dowex-50 (H⁺). After treating the homoarginine-containing column with water, to obtain another substance of interest, it was washed with 1 N HCl and developed with 1.5 N HCl. Arginine eluted sharply prior to the appearance of the rather diffuse homoarginine peak. The pooled homoarginine-containing eluent was exchanged again with the same Dowex resin, washed free of Cl⁻, and developed with 1 N NH₃; a yield of 2.8 gm (just under 0.1%) was secured. Homoarginine's identity was established by several chemical procedures including comparison of its flavianic and picric acid salts with authentic compounds. A comparable approach was taken by Chwalek and Przybylska (1970) in their isolation from *Lotus helleri*.

b. Biological Properties and Toxicity

Rao *et al.* (1963) tested the toxicity of this newly discovered compound and determined that 8 μg per ml culture medium causes a 50% reduction in *Streptococcus aureus* and *Candida albicans* growth; twice as much material prevented microbial proliferation. Earlier, Walker (1955a,b) used chemically synthesized material to characterize homoarginine not only as a potent inhibitor of *Chlorella vulgaris*—even more inhibitory than canavanine, but also as extremely toxic to *Escherichia coli*.

The primary route for mammalian production of guanidinoacetic acid, a creatine precursor, is by transamidination involving glycine and arginine:

L-Glycine + L-arginine → L-ornithine + guanidinoacetic acid

The substrate specificity of this transamidinase is not absolute (Ratner and Rochovansky, 1956) and lysine functions as the amidine group recipient with the resulting formation of homoarginine. It is not unexpected that homoarginine is synthesized by rat kidney which contains this enzyme (Ryan *et al.*, 1969).

As part of their study of α,γ-diaminobutyric acid in rats, O'Neal *et al.* (1968) also evaluated homoarginine and discovered that while 5 mmoles per kg body weight is innocuous, twice that amount produces hypersensitivity and death after 15 to 20 hr. Under conditions in which diaminobutyric acid significantly attenuates urea production, the arginine homolog does not alter sliced liver tissue function. The observation that 30 mM homoarginine does not curtail liver ornithine carbamoyltransferase activity is consistent with this finding. It is reasonable to suggest that a higher level of homoarginine is required to elicit adverse biological effects than is required for other toxic nonprotein amino acids. Homoarginine's limited toxicity has also been indicated by the experiments of D. L. Dahlman (personal communication) in which 25 mM homoarginine is injected into the hemolymph of fifth stadium tobacco hornworm larvae. Although pupae are produced from the treated larvae, they are deformed severely; no pupal malformations occur when the dose is reduced to 5 mM. At this dietary level, canavanine is very poisonous to this insect, and fifth stadium larvae consistently expire prior to larval–pupal ecdysis; any surviving pupae are deformed massively (Rosenthal and Dahlman, 1975). Other studies in Rosenthal's laboratory reveal that 10 μM homoarginine has no discernible effect on *Lemna minor* growth while this amount of canavanine prevents any frond proliferation.

Addition of 5% homoarginine to the diet of larval bruchid beetle, *Callosobruchus maculatus,* blocks adult emergence. When this compound is supplied in the diet at a substantially lower concentration, and one more in keeping with its typical concentration in nature (1% or less), it does not impede the appearance of the adult stage and it has a strong sparing effect on the action of canavanine (Janzen *et al.*, 1977). Consumption of this arginine analog by larval boll weevil, *Anthonomus grandis,* only slightly influences adult emergence and its effect is reversed readily by arginine; moreover, homoarginine can replace arginine in counteracting canavanine's adverse effects (Vanderzant and Chremos, 1971).

Thus, considerable variation exists in organism susceptibility to this arginine analog. While detailed studies of the biological effects of this compound have not been reported, it is reasonable to assume that it functions as a competitive inhibitor and antimetabolite of arginine.

5. γ-Hydroxy-L-arginine

At least 17 species of *Vicia* are able to synthesize this arginine derivative having a hydroxylated penultimate methylene group (Bell and Tirimanna, 1963).

$$H_2N—C(=NH)—NH—CH_2—CH_2—CH(OH)—CH_2—CH(NH_2)COOH$$
γ-Hydroxy-L-arginine

This compound has been obtained from *Polycheira rufescens*, a sea cucumber (Fujita, 1959), and shown subsequently to be a substrate for arginase, arginine decarboxylase, and L-amino acid oxidase (Fujita, 1961). Arginase-mediated hydrolysis of γ-hydroxyarginine proceeds at 11% of the reaction velocity attained with arginine; the relative value for its decarboxylation is 16.5%. It is synthesized by the sea anemone, *Anthopleura japonica*, along with such metabolic degradation products as β-hydroxy-γ-guanidinobutyric acid and γ-hydroxyagmatine (Makisumi, 1961). The latter compound represents a novel guanidino substance.

γ-Hydroxyarginine readily cyclizes, producing a lactone that has been obtained as a natural product of *Vicia sativus* (Bell and Tirimanna, 1964). Enzymatic hydrolysis of this amino acid would be expected to produce urea and γ-hydroxyornithine; the latter compound has, in fact, been isolated by partition paper chromatography and ionophoresis of an extract from *V. unijuga* and *V. onobrychoides* where it constitutes about 1% of the seed dry weight (Bell and Tirimanna, 1964).

6. γ-Hydroxy-L-homoarginine

In an early study of the free amino acids of *Lathyrus*, Bell (1962a) isolated two substances which he designated as B_1 and B_3 to denote their unknown but basic nature. B_1 was shown subsequently to be homoarginine while B_3 was identified as its γ-hydroxylated derivative (Bell, 1963a). Thus arginine, homoarginine, and their respective γ-hydroxy-containing derivatives are all part of the amino acid complement of *Vicia* and/or *Lathyrus*. As in the case of γ-hydroxyarginine, the higher homolog also cyclizes to form a lactone which has been isolated from *Lathyrus tingitanus* seeds (Bell, 1963b).

$$H_2N—C(=NH)—NH—CH_2—CH_2—CH(OH)—CH_2—CH(NH_2)COOH$$

γ-Hydroxyhomoarginine

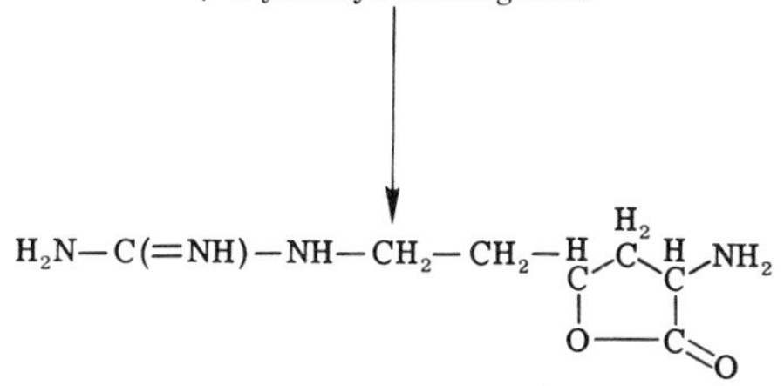

$$H_2N—C(=NH)—NH—CH_2—CH_2—H\underset{C}{\overset{H_2}{C}}H\underset{C}{\overset{}{}}NH_2$$

γ-Hydroxyhomoarginine lactone

A comparison of various chemical parameters of the synthesized dihydrochloride salt of the lactone to the natural product disclosed the absolute configuration to be *threo*-γ-hydroxy-L-homoarginine (Fugita *et al.*, 1965). This amino acid has also been reported in other legumes such as *Lens culinaris* (Wilding and Stahmann, 1962) and more recently in *Pisum sativum* (Smith and Best, 1976).

Studies of the toxicity of these hydroxylated arginine derivatives have not been reported and except for their relationship to the biosynthesis of lathyrine (considered in the next section) little is actually known of their role in higher plant intermediary metabolism. They were included here due to their structural similarity to arginine and the possibility that they may exhibit antimetabolic action against this protein amino acid.

7. L-Lathyrine

When routine chromatographic analyses of *Lathyrus* seeds revealed a compound forming a vivid orange-red color with ninhydrin, it became apparent immediately that this striking spot might portend a novel nonprotein amino acid. Since *L. tingitanus* produced the most intense color reaction, the newly isolated substance was sought from this source. It was obtained eventually as a crude sulfate salt from aqueous acetone, recrystallized from aqueous ethanol, and given the trivial name lathyrine (Bell, 1961).

The preliminary analytical chemistry led Bell to postulate its structure as that of a heterocyclic amino acid which opened upon hydrogenation to yield either a guanidino acid or substituted aminoguanidino acid. Further effort disclosed not only that upon reduction it absorbed two molecules of hydrogen, but also culminated in the isolation and characterization of the reduced derivative. In essence, these findings suggested the presence of a guanidino group as an integral part of a reduced ring system such as a substituted 2-aminopyrimidine. Nuclear magnetic resonance provided the critical data permitting its identification as β-(2-amino-pyrimidine-4-yl)alanine.

Isolation of this natural product was pursued simultaneously by several groups. The efforts of Nowacki and Przybylska (1961) culminated in the isolation and naming of the heterocyclic metabolite as tingitanine. Its concentration was given as 2.11% (dry weight) in *L. tingitanus* while other species of *Lathyrus* ranged from 0.22 to 0.52%.

When Bell ((1962c) isolated homoarginine, he appreciated immediately that it could represent a precursor of this heterocyclic amino acid, particularly since the natural occurrence and levels of the arginine homolog paralleled that of lathyrine (Bell, 1962a). More incisive, however, was the suggestion of Rao *et al.** (1963) that γ-hydroxyhomoarginine might be the imminent precursor compound. They predicted the eventual isolation of the requisite γ-hydroxyhomoarginine and its lactone from *L. tingitanus.*

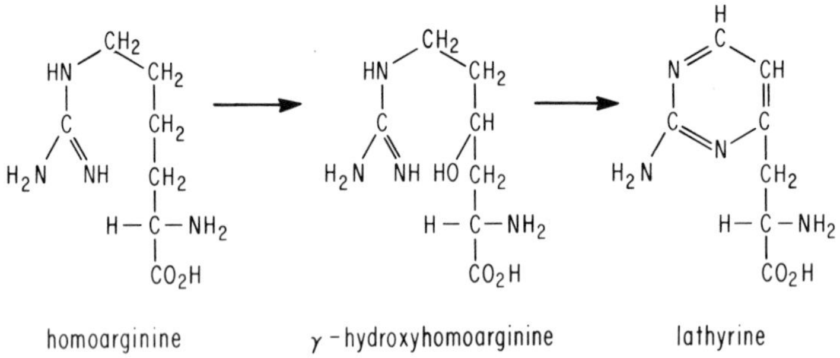

homoarginine   γ-hydroxyhomoarginine   lathyrine

Toward evaluating this putative pathway, L-[*guanidino-*[14]C]homoarginine was introduced into *L. tingitanus* either by injection into the pericarp or through the stem. Analysis of the seeds some 4 to 6 weeks later revealed a highly efficient transfer of the radioactive carbon to γ-hydroxyhomo-arginine. Little, if any, of the radioactivity was found either in other metabolically active amino acid pools or in lathyrine itself! The above reaction remained tenable if another tissue system, e.g., the roots, represented the actual site of γ-hydroxyhomoarginine conversion to lathyrine. When Bell and Przybylska (1965) fed γ-hydroxy-L-[*guanidino-*[14]C]arginine to the roots of this legume, two compounds of high specific activity were detected. The first was unreacted substrate but the other was lathyrine.

---

*These workers represented the third group independently pursuing the isolation of this heterocyclic compound since they were unaware of the efforts both in Poland and England.

Quite correctly, these workers repeated the initial experiment with radioactive homoarginine and root tissues and reported a strong labeling pattern in unreacted homoarginine, as well as hydroxyhomoarginine and lathyrine (Bell and Pryzybylska, 1965).

Further details of this reaction were proposed by Hider and John (1973) who contended the direct cyclization of γ-hydroxyhomoarginine was unlikely on chemical grounds and suggested instead the possible intermediate production of an oxo derivative that was converted to 5,6-dihydro-β-(2-aminopyrimid-4-yl)alanine and then lathyrine.

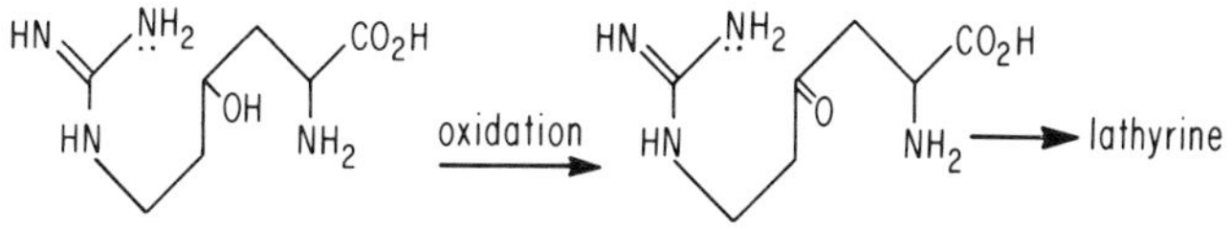

Quite recently, the matter of lathyrine biosynthesis was reopened as part of a study of pyrimidine ring synthesis (Brown and Al-Baldawi, 1977). Most pyrimidine constituents form in a reaction sequence involving orotic acid rather than cyclization of γ-hydroxyhomoarginine. Their most convincing findings supporting the existence and nature of an alternative pathway for producing lathyrine are the facts that [2-^{14}C]uracil specifically labels C-2 of lathyrine and that [2-^{14}C]uracil incorporation into lathyrine is greater than [6-^{14}C]orotate or even L-[guanidino-^{14}C]homoarginine. These workers proposed consequently the following bidirectional reaction scheme which incidentally does not recognize γ-oxohomoarginine's participation (Brown and Al-Baldawi, 1977):

Orotate → uracil → lathyrine ⇌ 5,6-dihydrolathyrine ⇌ γ-hydroxyhomoarginine ⇌ homoarginine.

D. SELENIUM-CONTAINING COMPOUNDS

a. Background and Constituents

Selenium is normally considered a minor soil component since it averages about 0.1 ppm (Olsen, 1978); under certain edaphic conditions, however, it accumulates in some higher plants well in excess of the supporting soil concentration. Of the accumulating plants, the best known are members of *Astragalus,* a genus of leguminous plants enjoying worldwide distribution. When the constituents of these selenium hoarders are analyzed, this element is found linked covalently to carbon, thereby forming structures wholly analogous to many common organic sulfur compounds (Shrift, 1969).

Approximately 500 species of *Astragalus* occur in North America but only two dozen members contain appreciable selenium even when growing on seleniferous soils (Shrift, 1967). Due both to their widespread distribution and intrinsic toxicity, this plant group while limited in number has nevertheless created significant range management problems over much of the western region of America. In actuality, all *Astragalus* growing on selenium-containing soils incorporate some of this element; what is pertinent is the degree and nature of the stored product. In sequestering plants, which can contain a staggering 15,000 ppm selenium, the predominant storage compounds are Se-methylselenocysteine* and selenocystathionine (Virupaksha and Shrift, 1965).

$$CH_3—Se—CH_2—CH(NH_2)COOH$$
Se-methyl-L-selenocysteine

$$HOOC—CH(NH_2)—CH_2—Se—CH_2—CH_2—CH(NH_2)COOH$$
L-Selenocystathionine

Se-methylselenomethionine, selenomethionine, and selenocystine are all characteristic compounds of the more ubiquitous non-accumulating species (Shrift, 1972). These distinctions are not absolute and non-accumulators can contain Se-methylselenocysteine and/or selenocystathionine but at a level less than that noted for accumulating plants (Martin *et al.*, 1971).

$$(CH_3)_2—Se^+—CH_2—CH_2—CH(NH_2)COOH$$
Se-methyl-L-selenomethionine

$$HOOC—CH(NH_2)—CH_2—Se—Se—CH_2—CH(NH_2)COOH$$
L-Selenocystine

The array of naturally occurring seleniferous amino acids reported in 1969 by Shrift is presented in Table 11.

b. Biological Properties

Lecythis ollaria, known commonly as coco de mono or monkey nut, is a deciduous species of Central and South America that produces a nutlike seed† whose consumption elicits abdominal pain, nausea, vomiting, and diarrhea; a week or two later, a reversible loss of scalp and body hair ensues (Aronow and Kerdel-Vegas, 1965). Drawing on a bioassay using

---

*A γ-glutamyl derivative exists and it is the principal selenium-containing compound of the seed of *Astragalus bisulcatus*. Hydrolysis of this seed dipeptide, thereby providing the free selenoamino acid, increases dramatically as the germinated seed grows (Nigam *et al.,* 1969).

†The nut is purported to contain up to 18,000 ppm selenium (Shrift, 1967).

Table 11 Selenium Compounds Reported in Higher Plants[a]

Selenium compound	Plant
Selenocystine	*Zea mays,* *Trifolium pratense, T. repens,* *Lolium perenne,* *Allium cepa,* *Spirodela oligorrhiza*
Selenocysteine– seleninic acid	*Trifolium pratense, T. repens,* *Lolium perenne*
Se-methylselenocysteine	*Astragalus* sp. *Oonopsis condensata, Stanleya* *pinnata*
Se-propenylseleno- cysteine selenoxide	*Allium cepa*
Selenohomocystine Selenomethionine	*Astragalus crotalariae* *Allium cepa,* *Trifolium pratense, T. repens,* *Lolium perenne,* *Spirodela oligorrhiza*
Se-methylselenomethionine	*Astragalus* sp. *Trifolium pratense, T. repens,* *Lolium perenne*
Selenomethionine selenoxide	*Trifolium pratense, T. repens,* *Lolium perenne*
Selenocystathionine	*Astragalus* species, *Stanleya pinnata,* *Lecythis ollaria,* *Neptunia amplexicaulis*

[a] Reproduced with permission from the *Annual Review of Plant Physiology*, Vol. 20 (1969). Copyright by Annual Reviews Inc.

cultured mouse fibroblasts, these workers instituted a search for the active factor in freshly collected nuts. Simultaneously, a research group at Wyeth Laboratories identified their crystalline biologically active material as selenocystathionine (Kerdel-Vegas *et al.*, 1965). The results of parallel testing of the biological and chemical properties of the toxic substance obtained by these two groups were identical. Essentially all of the cytotoxicity of the nut could be attributed to this selenium analog of cystathionine.

In a prior publication (1964), Francisco Kerdel-Vegas provided a fascinating accounting of an actual case history, which merits retelling:

H.D.P., a 54 year old Italian born white male, who had lived in Caracas for many years, consulted us May 5, 1962 because of extensive hair loss which he attributed to ingestion of "Coco de Mono". He stated that April 20, 1962 he ate 70 to 80 "Coco de Mono" almonds while on a hunting trip in the Venezuelan State of Portuguesa.

That same afternoon he felt nervousness and anxiety which was followed within 8 hours by violent chills. Subsequently he developed watery diarrhea consisting of 8 movements a day and persisting for 36 hours. After the fever and diarrhea subsided, there was a persistence of anorexia and asthenia. He developed arthralgia of all his joints, with pain in his chest and back. Anorexia, dyspepsia and a sensation of fullness has continued to the present time.

Eight days after having eaten the almonds there occurred a sudden, extensive loss of scalp and body hair which increased during the following days.

On May 3, 1962 he noted a violet streak over the proximal part of his nail plates which disappeared over the succeeding three days.

In the first few days he lost a great deal of hair from his axillae, chest, pubis and thighs. He also lost his eyebrows, eyelashes and mustache. Within a month the patient lost 4 kg. He gave a history of diuresis lasting one week after having eaten the fruit. He stated that the guide had warned him that his hair would fall out, when they found him eating the "Coco de Mono" almonds. On the same day that H. D. P. had ingested the almonds a companion also ate a moderate quantity of them (approximately 8 seeds), but did not note any ill effects. The guides stated that when monkeys eat this fruit they are found dead in the trees.

In a pioneering paper on animal selenium pathology, Rosenfeld and Beath (1964) classified selenium poisoning as being either acute, subacute, or chronic. In acute and subacute situations, the typical lesions consisted of multiple hemorrhaging and fibrosis of body organs. The chronic manifestations were subdivided into two maladies which have come to be known as "alkaline disease" and "blind staggers." In the former, inappetence, some visual disorientation, and aimless wandering are noted initially. Later, the front legs weaken and fail to provide adequate support for the animal. Hylin (1969) also includes depilation as well as deformity and sloughing of the hooves as part of the symptomology (Fig. 28). Selenomethionine and selenocysteine, provided by various non-accumulating plants growing on seleniferous soils such as forage grasses and cereals, are believed to be the causal agents of this malady.

A more severe form of selenosis of domesticated animals has been observed in two Australian regions. The legume *Neptunia amplexicaulis*, indigenous to one of these regions, is known to store a significant portion of its total selenium as selenocystathionine. The other region provides a habitat for *Morinda reticulata*, analysis of this plant revealed a selenium-containing compound representing 20% of the soluble amino acid nitrogen of the tops and containing 90% of the administered [75]Se. It

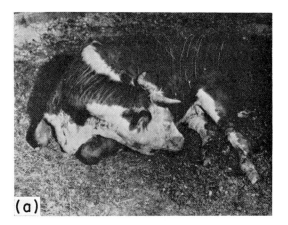

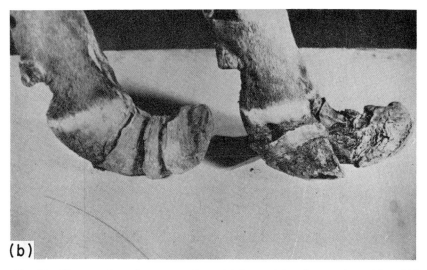

Fig. 28. Consumption of range weeds containing certain seleno amino acids, which are reputed to cause the symptoms of blind staggers (a). Other seleno amino acids of forage grasses and cereals inhabiting seleniferous soils cause alkaline disease, in which a deformity and sloughing of the hooves results (b). After Rosenfeld and Beath (1964).

was identified subsequently as selenocystathionine (Peterson and Butler, 1971).

Selenocystathionine and Se-methylselenocysteine, constituents of selenium-accumulating plants, are believed responsible for the acute selenium poisoning known as "blind staggers." The above compounds are ingested usually by range herbivores consuming such plants as members of the *Astragalus* and *Machaeranthera*. The resulting selenosis is most severe and debilitating (Fig. 28) and affected animals are seen to walk about aimlessly, to froth at the mouth, to exhibit indications of great physical pain, and to be subjected to potential respiratory failure within 24 hr (Hylin, 1969; Stadtman, 1974). Chronic poisoning in humans has also been reported from Columbia after grains cultivated on soils containing this toxic element were eaten (Rosenfeld and Beath, 1964).

On the other hand, trace levels of selenium (ca. 0.01 ppm) are very beneficial and this element is known to be essential for faunal development and growth. It is a component of glutathione peroxidase that mediates mammalian and avian reduction of organic peroxides. This protein possesses four monomeric units with a mass of 21,000 daltons, each probably containing 1 gm atom of selenium (see Stadtman, 1974). Selenium also functions in several prokaryotic enzymes (see Tanaka and Stadtman, 1979; Enoch and Lester, 1975).

c. Toxicity

In an informative review, Shrift (1972) has discussed the salient reasons for the early belief that selenium's toxicity simply reflected its property of sulfur antagonism. Convincing evidence was presented that the same enzyme complement actually functions or can function in parallel reaction pathways involving these somewhat chemically similar elements. Shrift hypothesized that the underlying reason for its toxicity was the substitution in protein of sulfur amino acids by their selenium-containing counterpart. Covalent intrapeptidyl bridges, such as C—S—Se—C or C—Se—Se—C, possess a lower energy barrier to rotation, thereby forcing a spatial orientation differing from that of a disulfide bridge. Moreover, when selenomethionine replaces methionine, the resulting protein's hydrophobic properties may be altered. These and other examples were presented and analyzed by Shrift (1972) by way of demonstrating how such residue replacement might affect ultimately various enzyme parameters. On the other hand, replacement of 70 to 75% of the methionine residues of *E. coli* β-galactosidase by selenomethionine failed to influence the catalytic activity. Overall protein synthesis, as assessed by monitoring [^{14}C]phenylalanine incorporation, was also unaltered. The only significant adverse effects reported were a 35% decrease in the rate of β-galactosidase induction and significant abatement of total protein synthesis when amino acids in addition to methionine limited prokaryotic growth (Coch and Greene, 1971).

A cysteinyl-tRNA synthetase capable of activating selenocysteine has been purified partially from mung bean (Shrift *et al.*, 1976) but direct experimental proof of selenocysteine in the protein of this legume is lacking (Burnell and Shrift, 1977).* Much more convincing, however, are the data supporting selenomethionine-containing protein production (Cowie and Cohen, 1957; Peterson and Butler, 1967).

Isolation of an active polysome fraction from *Vigna radiata* allowed Eustice *et al.* (1980) to demonstrate that selenomethionine and methionine were incorporated into protein with equal affinity. In contrast, when [^{75}Se]selenomethionine and [^{14}C]methionine were supplied concurrently, [^{75}Se]selenomethionine placement into protein lagged by 35% relative to [^{14}C]methionine. [^{75}Se]Selenomethionine labeled several high molecular weight proteins which, upon electrophoretic analysis, behaved identically to their [^{35}S]methionine-labeled counterpart (Eustice *et al.*, 1980).

Peterson and Butler (1967) developed another aspect of selenium biochemistry by their suggestion that accumulator species store selenium predominantly in their free nonprotein amino acid fraction while non-accumulators place most of this element into selenoamino acids that can function as protein constituents. Consistent with this assertion, callus cultures of non-accumulating species have relatively high selenium associated with the trichloroacetic acid-insoluble fraction while a lower selenium content is characteristic of these materials from accumulating plants (Ziebur and Shrift, 1971). It is not surprising that in accumulating species which can house 15 gm selenium per kg of tissue, this element becomes so pervasive and abundant that a preferred biological strategy is to produce selenoamino acids with limited ability to compete with their protein amino acid counterparts in the activation and aminoacylation processes.

Experiments with monogastric animals given [^{75}SeO$_3{}^{2-}$] suggested that these higher animals do not synthesize selenoamino acids (Jenkins, 1968). This finding is in accord with an earlier observation of Cummins and Martin (1967) that mammals are unable to produce these compounds. Jenkins conceded that selenium enters the protein pool but it is bound covalently between the sulfur atoms constituting half-cysteine residues, i.e., selenium is not brought directly into the macromolecule but rather via disulfide bond formation. If these workers are correct,

---

*This may reflect in part its lability to acid hydrolysis since it is nearly fully destroyed in 6 *N* HCl after 6 hr at 110°C (Huber and Criddle, 1967).

then the mere presence of radioactive selenium in the isolated protein of ^{75}Se-treated organisms is not proof of selenoamino acid production. Consistent with this important concept, Schwarz and Sweeney (1964) showed that inorganic selenium ions were able to bind to sulfur-containing amino acids.

The above results stand in contradistinction to a later study in which $[^{75}SeO_3{}^{2-}]$ was provided to lactating ewes; blood and milk samples were collected 48 hr later (Godwin *et al.*, 1971). These workers developed substantive chromatographic evidence that at least 3% of the labeled selenium was incorporated into milk proteins as bound selenomethionine. An examination of comparably treated rabbits demonstrated that while most inorganic selenium does not enter selenoamino acid pathways, the fraction that does is converted to selenocystine (Godwin and Fuss, 1972). It appears that animals may share with plants the ability to produce selenoamino acids. If so, selenomethionine and selenocystine are the predominant storage forms in these organisms. Selenium analogs such as choline selenate, flavinoid selenate, or selenoglutathione are not known to occur. In Shrift's words: "certain branches of sulfur metabolism are closed to selenium." The alternate is equally valid. *Astragalus vasei* fabricates Se-methylselenocysteine from selenite but cannot form S-methylcysteine from [35]S.* *Astragalus crotalariae* processes ^{35}S-methionine essentially unchanged while ^{75}Se-selenomethionine yields Se-methyl-selenomethionine, selenohomocystine, and Se-methylselenocysteine (Shrift, 1967).

d. Metabolism

Experiments involving labeled precursors and *Astragalus bisulcatus* culminated in the following putative reaction scheme for Se-methylselenocysteine formation (Chen *et al.*, 1970):

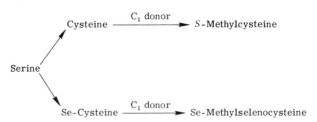

---

*This is an unusual situation, since these two forms of cysteine almost always coexist in the plant (Martin *et al.*, 1971).

Methionine was viewed as the C_1 donor for the transmethylation reaction. Some evidence of selenomethylselenocysteine production at the expense of its sulfur amino acid counterpart was offered, i.e., these reactions compete for serine. It has also been suggested that Se-methylselenocysteine can be produced from selenomethionine (Chow *et al.*, 1972). This can be achieved by selenomethyl group transfer from selenomethionine to serine or by a transsulfuration-type reaction involving selenocystathionine; the latter alternative was favored by these workers (Chow *et al.*, 1973; see Chapter 4, Section B,1).

Stanleya pinnata is able to synthesize selenocystathionine from selenate (Virupaksha and Shrift, 1963). A more important finding of this study was their observation that several established selenium accumulators, i.e., *A. crotalariae*, *A. bisulcatus*, and *Oonopsis condensata* do not store selenocystathionine which is most intriguing in view of its postulated role in Se-methylselenocysteine formation. It is germane that two isomeric forms of the γ-glutamyl derivative of cystathionine have been isolated from *Astragalus pectinatus* (Nigam and McConnell, 1976); this finding makes the natural occurrence of cystathionine in *Astragalus* much more certain.

Garden variety cabbage *Brassica oleraceae capitata* is a facile producer of selenoamino acids. Foliage from plants that had been grown for two months on soil fertilized with [$^{75}SeO_3^{2-}$]contained the following labeled metabolites: selenomethionine, Se-methylselenomethionine, Se-methylselenocysteine, Se-methylselenocystine, selenohomocystine, and selenocystathionine (Hamilton, 1975). An enzyme has been isolated from this plant that converts the selenonium salt of Se-methylselenomethionine to dimethyl selenide and homoserine. This finding is most significant since it integrates the metabolic reactions of selenoamino acids with that of amino acid carbon and nitrogen metabolisms (Lewis *et al.*, 1971).

In her review of selenium in higher plants, Lewis (1976) proposed the following scheme for the selenium metabolism of non-accumulating higher plants:

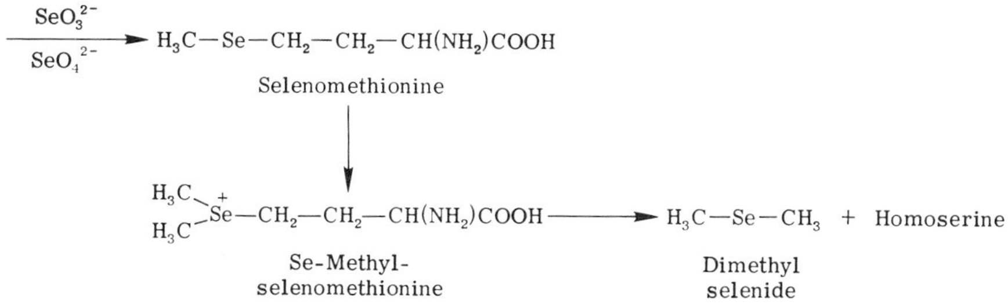

$$\frac{SeO_3^{2-}}{SeO_4^{2-}} \longrightarrow H_3C-Se-CH_2-CH_2-CH(NH_2)COOH$$

Selenomethionine

$$H_3C{\underset{H_3C}{\overset{+}{\diagdown}}}Se-CH_2-CH_2-CH(NH_2)COOH \longrightarrow H_3C-Se-CH_3 \; + \; \text{Homoserine}$$

Se-Methyl-
selenomethionine   Dimethyl
selenide

This scheme reflects the finding that selenomethionine and Se-methylselenomethionine are prime selenium storage forms of non-accumulating species. On the other hand, accumulators adhere to this scheme:

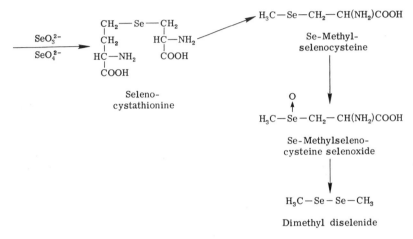

Again, this scheme reflects the presence of selenocystathionine and Se-methylselenocysteine in selenium accumulators. The final two components of this scheme are largely inferential in nature, being derived from comparable reactions of sulfur metabolism. Dimethyl diselenide, however, is a volatile component of *A. racemosus* (Evans *et al.*, 1968).

E. MISCELLANEOUS COMPOUNDS

1. L-Hypoglycin

a. Background and Isolation

The tree *Blighia sapida* produces an arillus (tissue covering the seed) which, on a dry weight basis, stores about $0.008 \pm 0.001\%$ of the toxic nonprotein amino acid, hypoglycin A (Hassall and Reyle, 1955).

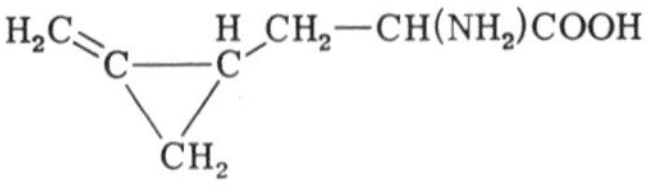

$$H_2C{>}C{-}C{<}{}^{H}_{CH_2}{-}CH(NH_2)COOH$$
$$\diagdown CH_2 \diagup$$

L-Hypoglycin

Indigenous to coastal West Africa where it is known as "ishin," this tree was introduced into Jamaica in 1778. Eventually, the "ackee tree," as it is known locally, attained a distributional pattern covering the New World tropics (Hill, 1952). Ripe arilli are eaten throughout Jamaica where their consumption is *associated* with a disease known as "vomiting sickness" (Scott, 1917). The characteristic syndrome includes a violent retching and vomiting that inspired its name. Vomiting is followed usually by convulsions and coma; death can occur within 2 or 3 days. The symptoms manifest rapidly, often within a few hours, and they are particularly severe in children and malnourished adults (Ellington, 1976).

Reports of the incidents of vomiting sickness reveal its strong seasonal aspect, which can reach epidemic levels during the period of the year when plants are dormant. At that time, food is scarcer and unripened ackee seeds become an increasingly attractive and utilized food resource. This situation is further exacerbated by the greater hypoglycin A content of the unripened arilli, some $0.111 \pm 0.005\%$ (Hassall and Reyle, 1955). A fascinating first-hand account of a physician's experience with the incidence and treatment of "vomiting sickness" has been prepared by Stuart (1976).

A γ-glutamyl peptide of hypoglycin A is also produced by *Blighia* (Hassall and Reyle, 1955) and it is found in both the unripened arilli and especially the seed (Fowden, 1976). Hassall and John (1960) have suggested that hypoglycin A be referred to simply as hypoglycin while hypoglycin B, the peptidyl derivative, be designated γ-glutamyl-L-hypoglycin. Since this suggestion is more in keeping with informative nomenclature, it has been adopted throughout the remainder of this work.

Hypoglycin and γ-glutamyl hypoglycin have now been isolated from other species of *Blighia* and *Billia* (a genus of Hippocastanaceae) and *Acer pseudoplatanus* (Eloff and Fowden, 1970; Fowden and Pratt, 1973). A lower homolog of hypoglycin, namely α-(methylenecyclopropyl)glycine, has been identified as a natural product of *Litchi chinensis*, another sapindaceous species (Gray and Fowden, 1962); this compound also exhibits hypoglycemic properties.

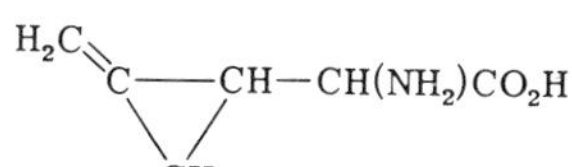

α-(Methylene-
cyclopropyl)glycine

Seeds of *Billia* (Eloff and Fowden, 1970) and *Acer* (Fowden *et al.*, 1972b) store α-(methylenecyclopropyl)glycine and its corresponding γ-glutamyl peptide, therefore a lower homologue of γ-glutamyl hypoglycin, also occurs in the latter plant.

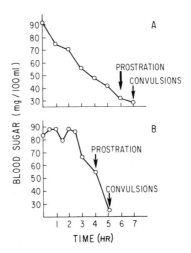

Fig. 29.   Induction of hypoglycemia. The treated rabbit received 20 mg hypoglycin (A) or 25 mg γ-glutamylhypoglycin (B) per kilogram body weight by intravenous administration. Reprinted with permission from Chen *et al.* (1951). Copyright by Williams & Wilkins Company.

b.   Biological Properties and Toxicity

One of the most striking clinical manifestations of hypoglycin consumption is hypoglycemia, a condition in humans in which blood glucose level can fall to an astounding 3 mg%* (Bressler *et al.*, 1969) (Fig. 29). Yet, glucose catabolism itself is little affected, as judged from its unimpaired conversion to respiratory CO_2, liver lipid, or production of muscle glycogen (deRenzo *et al.*, 1958). Hepatic glycogen reserves can be so severely depleted that a single intramuscular injection of hypoglycin almost completely exhausts this energy-reserve polymer (Patrick, 1954). Fatty infiltration or metamorphosis of the liver and other organs, with a concomitant enrichment in hepatic lipids, is a common postmortem observation of poisoned animals (von Holt and von Holt, 1958). The mitochondria swell and an unusual central juxtapositioning of the cristae of certain hypoglycin-treated animals often ensues (Fig. 30). High-energy phosphate bond production by isolated liver mitochondria is impaired, especially in association with pyruvate and malate oxidations (McKerns *et al.*, 1960). Hypoglycin administration does not affect [U-^{14}C]glucose or [1-^{14}C]acetate conversion to respiratory CO_2 but it does attenuate respiration of [1-^{14}C]butyrate, -palmitate, and -stearate to CO_2 (McKerns *et al.*, 1960).

*Human blood normally contains around 100 mg% glucose.

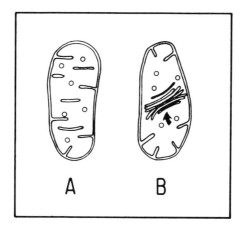

Fig. 30. Hepatocyte mitochondria. The organelle depicted in (A) represents material obtained from an untreated rat, while sample (B) was derived from a rat given 1 mg hypoglycin/100 gm body weight for 10 days and fasted on the final day. The arrow indicates the stacked central cristae pattern. After Brooks and Audretsch (1976).

From the above experimental observations and others, a consensus has emerged that the toxicity of this ringed natural product resides in the realm of fatty acid oxidation. Studies reported by von Holt *et al.* (1966) suggest that mitochondrial-mediated oxidation of long chain fatty acids (C_{12}–C_{14}) is prevented. Fatty acids are activated normally by various acylating agents that produce a variety of acyl-CoA derivatives. These acyl-CoA molecules are transported to the mitochondria where they are dehydrogenated by a flavin-requiring dehydrogenase. It is this dehydrogenation which von Holt *et al.* (1966) contend is affected by hypoglycin.

On the other hand, the work of McKerns *et al.* (1960) and Brendel *et al.* (1969) indicate that short-chain fatty acid oxidation is affected adversely by hypoglycin, providing the treatment duration is long enough for this toxic amino acid to work. While the precise details have yet to emerge, there is ample experimental evidence that hypoglycin obstructs fatty acid oxidation and that cellular gluconeogenesis is curtailed as well. Abated fatty acid oxidation places an additional burden on glucose reserves to meet the organism's energy needs. Since gluconeogenesis is also diminished, it is inevitable that glucose reserves become exhausted and hypoglycemic symptoms ensue. This rationale was first promulgated explicitly by Senior (1967) to explain the hypoglycemic action of pent-4-enoic acid.

A hypoglycemic compound characteristically possesses a C to C double bond separated from an actual or potential carboxyl group by two carbon atoms (Corredor, 1976). Pent-4-enoic acid is a relatively simple compound that complies with these structural requirements. It has become an accepted alternative to hypoglycin in pharmacological studies since they are believed to share a common mechanism of action. This factor obviates the relative scarcity of the natural product, the experimental difficulties associated with its isolation, and its frequent vitiation with leucine or other natural products (it is difficult but possible to isolate homogeneous hypoglycin from biological sources, see p. 40).

The precipitous decline in blood glucose level resulting from hypoglycin consumption usually occurs only after 3 or 4 hr. This delay may represent the time frame for hypoglycin conversion to the actual toxic principle (McKerns *et al.*, 1960). That such a metabolically active constituent exists and may, in fact, be methylenecyclopropylacetic acid has been suggested by several independent investigations. This putative active factor is a potent inhibitor of fatty acid oxidation (von Holt *et al.*, 1966).

The reactions responsible for this metabolic conversion include transamination of hypoglycin to α-methylenecyclopropylpyruvic acid, oxidative decarboxylation and formation of the CoA ester of α-methylenecyclopropylacetate, and deacylation to the reputed active compound (Fig. 31) (Sherratt and Osmundsen, 1976).

Bressler *et al.* (1969) have integrated many of these experimental findings and concluded that hypoglycin, or a toxic metabolic product, is activated to form acetyl-CoA derivatives that compete for carnitine acetyltransferase. This depresses intracellular carnitine and coenzyme A levels, thereby curtailing fatty acid oxidation. Consistent with this belief, Entman and Bressler (1967) demonstrated that carnitine, which stimulates mitochondria-dependent fatty acid oxidation, spares the hypoglycemic action of hypoglycin. In essence, inhibited fatty acid oxidation

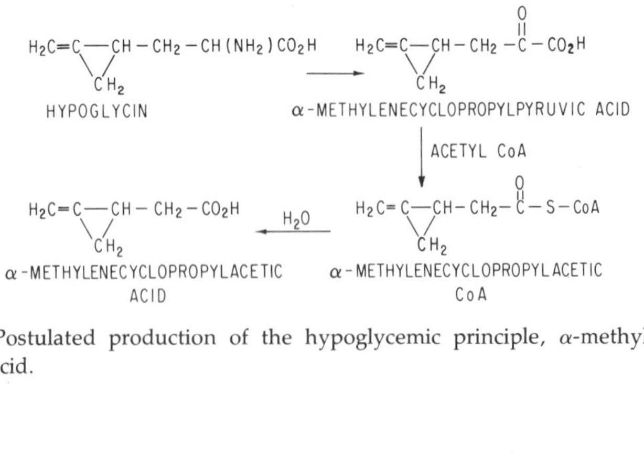

Fig. 31. Postulated production of the hypoglycemic principle, α-methylenecyclopropylacetic acid.

and hypoglycin-dependent disruption of mitochondrial high energy phosphate production combine to reduce acetyl-CoA, ATP, and NADH levels below those required for optimal gluconeogenesis. Since fatty acid oxidation products regulate glycolytic reaction rates, their loss further exacerbates glucose depletion by accelerating glycolytic activity.

In an unwarrantedly critical analysis of the concepts of Bressler *et al.*, Sherratt and Osmundsen (1976) nevertheless presented the cogent argument that short chain acyl-CoA hydrolases provide a means for compensatory CoA production from cellular acyl-CoA pools. They give little credence to a central thesis of Bressler *et al.* (1969) that hypoglycin can cause significant loss of CoA. Sherratt and Osmundsen were unable to reproduce those experiments of Entmann and Bressler indicating that carnitine can spare the hypoglycemic action of hypoglycin. Among other reasonable arguments, they rejected the notion of CoA sequestration because of the previously discussed structural requirement for a given compound to elicit hypoglycemic properties. Sherratt and Osmundsen believe that this cyclic compound and other related pharmacologically active substances must be specific inhibitors of enzyme(s) functioning in β-oxidation. These workers point out that dihydrohypoglycin is not hypoglycemic although this derivative would be metabolized to methylcyclopropylacetyl-CoA, which they contend should be equally effective as methylenecyclopropylacetate CoA in complexing free CoA. These investigators prefer the explanation that hypoglycin or pent-4-enoate functions by inhibiting gluconeogenesis and subsequently fatty acid oxidation through an active form which specifically and reversibly inhibits enzymatic reaction(s) of β-oxidation. It is evident that additional work must be completed to fully clarify the exact mode of action of this interesting natural product.

Hypoglycin is also highly teratogenic in rats (Persaud, 1967) but not in chicks (Persaud and Kaplan, 1970). In rats, leucine fails to spare its teratogenicity, while significantly amplifying fetal resorption (hypoglycin is believed to be a metabolic antagonist of leucine). Riboflavin phosphate, in contrast, attenuates fetal abnormalities (Persaud, 1970). The latter experimental finding is in keeping with the earlier discussed suggestion of von Holt *et al.* (1966) that hypoglycin's site of action is at the level of flavin-dependent acyl dehydrogenase, providing a relationship exists between acyl-CoA formation and teratogenicity.

γ-Glutamyl hypoglycin also elicits teratogenic effects in rats. Simultaneous administration of γ-glutamyl hypoglycin and riboflavin intensifies fetal abnormality as compared to riboflavin omission. In this instance, leucine can assuage the fetal aberrations caused by this toxin (Persaud, 1968). These intriguing dichotomies of experimental action forced Persaud (1976) to propose independent mechanisms of terato-genicity for hypoglycin and its peptidyl derivative.

Protein synthesis obviously plays a critical role in embryonic and fetal development. Persaud has suggested that an imbalance in amino acid availability either reflecting a diminution in leucine content or some other factor adversely affects fetal development through curtailed protein synthesis. However, no experimental evidence was provided to support this speculation. It would be worthwhile to explore the biochemical basis for the divergent results obtained with this nonprotein amino acid and its γ-glutamyl derivative.

2. L-Azetidine-2-carboxylic Acid

a. *Background and Isolation*

Azetidine-2-carboxylic acid was the first nonprotein amino acid discovered to have the relatively simple four-atom azetidine ringed structure although several related natural products are now known.

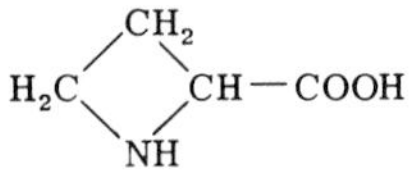

L-Azetidine-2-carboxylic acid

It was isolated initially by Fowden (1955) from an aqueous ethanolic extract of *Convallaria majalis*. His reconstruction of the employed analytical methodology is an illustrative example of a systematic approach to natural product isolation. Virtanen and Linko (1955) achieved a concurrent isolation from *Polygonatum officinale*. The early sources of this amino acid included members of the Liliaceae as well as representatives of the Agavaceae, a closely related plant group. These plants accumulate 3 to 6% of their dry weight as azetidine-2-carboxylic acid but it constitutes 75% of the nonprotein amino acid nitrogen contained in the storage organ (rhizome) of *Polygonatum multiflorum*.

Fowden and Steward's 1957 survey of 90 species of various liliaceous genera demonstrated the presence of azetidine-2-carboxylic acid in about one-quarter of the analyzed plants. The isolation of this compound from the legume *Delonix regia* was unanticipated since it had been assumed to be associated uniquely with liliaceous plants (Sung and Fowden, 1969). It has been secured subsequently from other leguminous species including *Parkinsonia aculeate*, *Busse massaiensis*, *Schizolobium parahybum*, *Peltophorum inerme*, and *P. africanum* (Watson and Fowden, 1973). That these distinctive plant groups were shown to be sources of this compound after it had long been believed to be associated uniquely with liliaceous plants highlights an important consideration. Just how cosmopolitan would the distribution of any nonprotein amino acid be, if the sensitivity of present analytical methodology was amplified significantly and many diverse plant groups were analyzed systematically for its occurrence? This question bears strongly on the present practice of using particular nonprotein amino acids or other secondary metabolites to delineate higher plant phylogenetic relationships (see Fowden, 1972).

b. Biological Properties and Toxicity

Few studies of the biological effects of this amino acid have been conducted. No information is available on whole animal effects in mammals but this compound causes developmental aberrations in chick embryos and the egg of *Paracentrotus lividus*, a sea urchin (Lallier, 1965). The liliaceous plants *Ruscus hypoglossum* and particularly *Urginea maritima* are resistant to the lepidopteran predator, *Spodoptera littoralis*. *Urginea maritima* contains 1.7% azetidine-2-carboxylic acid on a fresh weight basis and this accounts for 85% of the total free amino acid pool (Hassid *et al.*, 1976). After extracting the foliage with 80% aqueous methanol, the soluble components were processed to isolate the pharmacologically active components. Ultimately, the applied analytical procedures established convincingly that foliar azetidine-2-carboxylic acid was the primary cause of the mortality observed in various test larval stadia.

Lemna minor and *L. gibba* can be sustained on a medium containing glutamine as the sole nitrogen source but when the fronds are transferred to an ammonium-containing medium, NAD^+-dependent glutamic acid dehydrogenase activity increases (Joy, 1969). This phenomenon is reported to result from *de novo* enzyme synthesis on 80 S ribosomes. Azetidine-2-carboxylic acid and the artificial nonprotein amino acid, *p*-fluorophenylalanine, not only completely terminate synthesis of this enzyme, but also continue to diminish the enzyme activity existing at the time of frond transfer to the ammonia-containing medium (Shepard and Thurman, 1973). These workers also showed that this proline analog, given at a 1 mM level, reduces total respiration by two-thirds.

When azetidine-2-carboxylic acid is provided (100 μg ml^{-1}) to exponentially growing *E. coli*, the shape of the resulting growth inhibition curve mimicked that noted for systems in which an amino acid analog is incorporated into protein (Fowden and Richmond, 1963).

3,4-Dehydroproline is also a strong microbial and higher plant growth retardant and this proline antagonist is incorporated into protein. In contrast, 4,5-dehydropipecolic acid (baikiain) and pipecolic acid, structurally similar to proline, are neither potent growth inhibitors nor do they appear to be incorporated into protein (Peterson and Fowden, 1965). The C to C bond of proline is 1.54 Å which is reasonably close to the 1.33 Å size of the $C=C$ bond constituting 3,4-dehydroproline. 4,5-Dehydropipecolic acid is much smaller and pipecolic acid much larger than proline (Fowden *et al.*, 1967). The size characteristics of these compounds offer a plausible explanation for their inability to compete with proline in amino acid activation by prolyl-tRNA synthetase and their resulting lack of antimetabolic action.

An investigation particularly relevant to this point was instituted by Peterson and Fowden (1965) who isolated the prolyl-tRNA synthetase from mung bean as well as from the rhizome of *Polygonatum multiflorum*. The bean enzyme, obtained from a source free of azetidine-2-carboxylic acid, activates this amino acid analog at about one-third the rate observed from proline. However, when the enzyme is secured from the rhizome producing azetidine-2-carboxylic acid, it fails to activate the nonprotein amino acid. Similarly, the bean enzyme stimulates ATP-PP$_i$ exchange with DL-3,4-dehydroproline at 57% of the rate obtained with proline but this value falls to 28% when the prolyl-tRNA synthetase is obtained from *P. multiflorum*.

These experiments are noteworthy in indicating for the first time the contribution of amino acid analogue activation and the concomitant formation of anomalous protein, in accounting for the toxicity of these secondary plant metabolites. In this regard, it is also known that the azetidinium ring bond angle turns the α-helix of the polypeptide chain through an angle of 15° less than that resulting from the pyrrolidine ring of proline (see Fowden, 1963); this can have a profound effect on the tertiary structure of the resulting protein.

Azetidine-2-carboxylic acid is a very potent antagonist of [^{14}C]proline assimilation into the collagen of embryonic cartilage, apparently by competing with the uptake of proline (Takeuchi and Prockop, 1969). It is also incorporated into developing chick embryonic protocollagen where it obstructs posttranslational hydroxylation of both proline and lysine residues. In these various ways, it impedes hydroxylation of protocollagen, a requisite to normal collagen formation; decreases the rate at which collagen is extruded and subsequently accumulated; and enhances the overall fragility of the embryonic cartilage (Lane *et al.*, 1971). (Compare this with the action of mimosine, see p. 84).

c. *Metabolism*

Radioactive precursor feeding studies by Fowden and Bryant (1959) failed to reveal appreciable labeling of this amino acid from such potential precursors as aspartic or glutamic acids and other putative intermediates. In contrast, Leete (1964) reported that 1.67% of the labeled carbon provided as DL-[1-^{14}C]methionine was recovered in azetidine-2-carboxylic acid. He proposed that displacement of thiomethyladenosine by the α-amino group of S-adenosylmethionine culminated in the biosynthesis of this nonprotein amino acid:

$$\underset{\substack{\uparrow\\NH_2-CH-COOH}}{CH_3-\overset{\overset{\displaystyle Adenosine}{|}}{S^+}-CH_2-CH_2} \longrightarrow CH_3-\overset{\overset{\displaystyle Adenosine}{|}}{S} \quad + \quad \underset{\substack{|\\HN-CH-COOH}}{\overset{\substack{H_2C-CH_2\\| \quad\quad |}}{}}$$

When DL-[*carboxy*-^{14}C]methionine was fed to *Convallaria majalis,* nearly all the radioactivity transferred to azetidine-2-carboxylic acid was recovered from its carboxyl group. This was determined by heating with ninhydrin and trapping the evolved $^{14}CO_2$ as barium carbonate. This study only revealed a possible metabolic relationship between methionine and azetidine-2-carboxylic acid since no direct evidence was provided for a functional role for the S-adenosyl derivative (see Su and Levenberg, 1967).

Utilizing labeled precursor feeding techniques, Sung and Fowden (1971) observed that the carbon atoms of homoserine and 2,4-diaminobutyric acid contributed radioactivity to the imino acid more effectively than did methionine. This occurred in spite of the fact that only the latter compound was provided as the racemically pure L enantiomer (this factor lacks significance if an active racemase operated in this system). This observation led Sung and Fowden to disfavor the S-adenosyl derivative of methionine as the *in vivo* precursor of this imino acid and to favor instead a C_4-chain contribution from another progenitor such as homoserine or 2,4-diaminobutyric acid. Methionine was not precluded as a reaction intermediate but the means for cleaving the sulfur bond as originally outlined by Leete was considered invalid.

Ten years later, Leete *et al.* (1974) reopened this question and finally discounted methionine and its S-adenosyl derivative as *immediate* precursors. These workers contended instead that 2-oxo-4-aminobutyric acid cyclized to form 1-azetine-2-carboxylic acid. They assimilated the findings of Sung and Fowden by proposing 2,4-diaminobutyric acid as a reaction intermediate via 2 oxo-4-aminobutyric acid. Since homoserine forms this diamino acid in *Lathyrus sylvestris* (Nigam and Ressler, 1966), it was designated a putative reaction component. They held tenaciously to a postulated role for S-adenosylmethionine by suggesting a possible amination reaction as part of its involvement in azetidine-2-carboxylic acid formation. These relationships are summarized in Fig. 32.

Azetidine-2-carboxylic acid catabolism has been investigated in the soil-inhabiting bacterium *Agrobacterium spp.* (Dunnill and Fowden, 1965b). While their findings may not be applicable to degradative path-

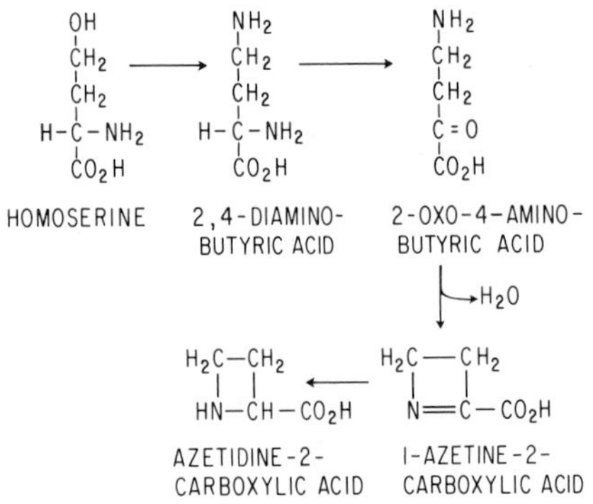

Fig. 32. A possible route for the generation of azetidine-2-carboxylic acid. Adapted from the work of Leete *et al.* (1974).

ways in eukaryotes, they are not only intrinsically interesting, but also highlight the frequent interconversion occurring between reactions of primary and secondary metabolism.

In its degradation in the soil, azetidine-2-carboxylic acid (I) may be oxidized (II) and the ring spontaneously opened to yield 2-oxo-4-amino-butyric acid (III). Subsequent reduction yields 2-hydroxy-4-aminobutyric acid (IV) (Fig. 33). Alternatively, this process can be initiated by hydrolytic ring cleavage to produce (IV). In either event, 2-hydroxy-4-aminobutyric acid would function as a nitrogen donor for transamination of

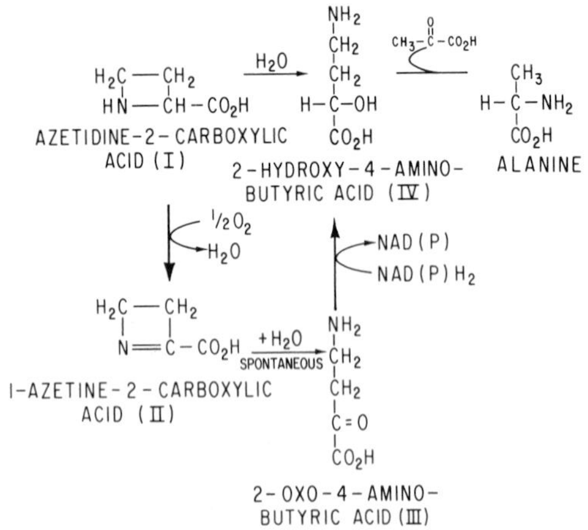

Fig. 33. Decomposition of azetidine-2-carboxylic acid by a soil-borne bacterium, *Agrobacterium* spp.

pyruvate to alanine with malic-γ-semialdehyde as a reaction product or intermediate if malic acid is eventually formed. Glutamic acid cannot be produced from 2-oxoglutaric acid nor is proline or pipecolic acid transformed by these pathways. However, other α-amino group-containing compounds function as donor species.

3. L-Djenkolic Acid

The djenkol bean has several sources but it is best known from *Pithecolobium lobatum,* a leguminous tree of Java, which commits about 1 to 2% of its dry seed products to a structurally unusual nonprotein amino acid known as djenkolic acid.

$$HOOC—CH(NH_2)—CH_2—S—CH_2—S—CH_2—CH(NH_2)COOH$$

L-Djenkolic acid

This natural product is more prevalent, as much as 4% of the dry weight, in the seeds of a companion species from Sumatra, *P. bubalinum* (van Veen and Latuasan, 1949). Chromatographic analyses of an extract of dried young leaves and twigs failed to produce a "positive" spot for this compound. While it may be absent from these tissues, a more probable explanation is that it is labile and simply lost during the drying process. Djenkolic acid has also been reported in certain members of the Mimosaceae and its N-acetylated derivative from various *Acacia* and *Mimosa acanthocarpa* (Gmelin *et al.,* 1962).

Djenkolic acid, sparsely soluble under acidic conditions, readily crystallizes from the urine of individuals consuming sufficient quantity of the seed. This can cause acute kidney malfunction which can ramify into anuria, i.e., impaired or blocked urine flow (van Veen, 1973). In young children crystal formation has been reported to obstruct the ureter and produce painful swelling of the external genitalia (Suharjono and Sadatun, 1968). These workers reported a higher incidence of such symptoms between September and January when the blossoms make their appearance. The floral structures may act as an attractant for children who then eat the immature seed which can, like hypoglycin, contain a higher content of the toxic principle.

In spite of its known toxicity, Indonesians prize the bean, possibly relishing its sour taste. The disagreeable odor, by Western standards, is of little consequence and does not mitigate against its widespread consumption, particularly by residents of Western Java who enjoy a better overall diet than their fellow countrymen. It may serve as a substitute for other amino acids in providing dietary sulfur but this point is pure conjecture (Hijman and van Veen, 1936).

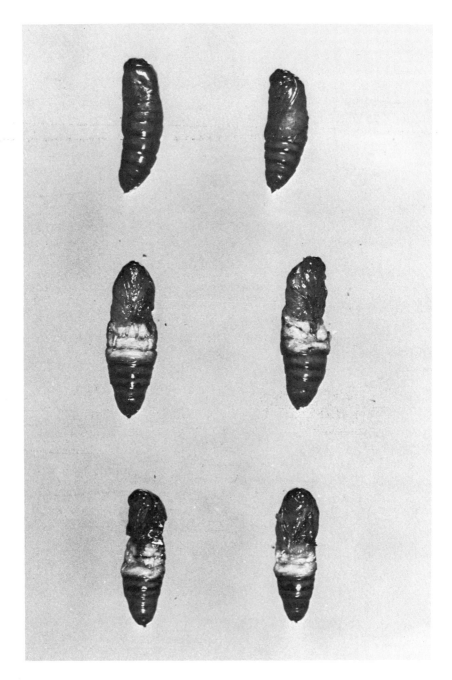

Fig. 34. *Prodenia eridania* pupae from L-dopa-treated larvae. The dramatic unsclerotized pupal cases result from larval consumption of a 5% bean leaf powder diet supplemented with 0.25% L-dopa (bottom row) or 5% *Mucuna pruriens* seed powder

F. TOXICITY STUDIES

The relative toxicity of some nonprotein amino acids has been evaluated in general feeding studies with insects and phytotoxicity determinations with higher plants. While limited in scope, they neverthe-. less represent the known data base for comparative examination of these toxic natural products. One of the earliest insect tests was conducted with the southern armyworm, *Prodenia eridania,* which was provided an artificial diet containing the test compound as a percentage of the total diet weight (Rehr *et al.,* 1973a). Canavanine (5%), 5-hydroxytryptophan (8%) and β-hydroxy-γ-methylglutamic acid (4%) caused total larval mortality within 3 days. Addition of 3.2% S-(β-carboxyethyl)-cysteine or low levels of L-dopa (ca. 0.25%) produced moderate larval deformity. L-Albizziine was found to have little potency with this insect.

In another study with the polyphagous feeder *Prodenia eridania,* L-dopa was incorporated into the artificial diet used to rear the larvae. All larvae exposed to 5% L-dopa ceased feeding and rapidly expired; this concentration approximates that found in the *Mucuna* seed. Pupal and adult malformations were typical of individuals developing from diet containing 5% *Mucuna* seed powder. The pupal case of the L-dopa-treated animals failed to sclerotize over the ventral head, thoracic, and first three abdominal segments (Fig. 34). Tyrosinase is known to contribute to the hardening and darkening of the cuticle of insects. These authors suggest that L-dopa or a metabolic derivative may affect normal aging of insectan cuticle and sclerotization of the adult (Rehr *et al.,* 1973b).

The insecticidal potential of some nonprotein amino acids has also been evaluated by injection into the hemolymph of newly ecdysed fifth stadium tobacco hornworm, *Manduca sexta* (D. L. Dahlman and G. A. Rosenthal, unpublished observations). At a level of 8 μmoles/gm fresh body weight, canavanine, canaline, mimosine, selenomethionine, and especially methionine sulfoximine were exceedingly toxic while α,γ-diaminobutyric acid and β-cyanoalanine were less toxic. At this concentration, L-dopa, azetidine-2-carboxylic acid, and albizziine were not effective in reducing growth.

Janzen *et al.* (1977) incorporated the test compound into ground cowpea (*Vigna unguicultata*) seed powder and then "pressed" the mass into tablets that functioned as surrogate seeds for egg deposition and

---

supplemented diet (middle row); the top row represents the control specimens obtained from larvae maintained on 5% powder from bean leaves. Photo kindly supplied by P. Feeny. Taken from the work of Rehr *et al.* (1973b). Copyright by the American Association for the Advancement of Science.

development. The number of emerging beetles was subsequently re-corded. All treated bruchid beetles (*Callosobruchus maculatus*) died when reared on tablets containing 5% (w/v) of any of the test non-protein amino acids except homoarginine and racemic pipecolic acid. At this high dose, however, many protein amino acids are also toxic and it may be that this organism dies simply because it cannot adequately eliminate so much dietary nitrogen. When the tested concentration decreased to 1.0% or less (e.g., 0.1%), emergence increased drastically.

The approach of Navon and Bernays (1978) consisted of placing sucrose solutions containing the test compound onto glass fiber discs which were dried and offered to three test organisms. The first was the migratory locust, *Locusta migratoria migratorioides*, a ravenous graminivorous feeder. Second was *Chortoicetes terminifera*, a less restricted feeder. The final test species was a polyphagous organism, *Schistocera americana gregaria*. The polyphagous feeder *Schistocera*, perhaps reflecting its enhanced detoxification capability, most effectively tolerated the tested natural products. L-Dopa (1 mM) was the most potent compound examined; at a 10 mM level, canavanine and albizziine exhibited significant toxicosis. A greater dose, i.e., 100 mM*, was required for the remaining nonprotein amino acids to retard insect feeding activity. This report by Navon and Bernays is also of interest for it represents the first indication of the sensitivity of an insect to albizziine. As such it reinforces the relative nature of nonprotein amino acid toxicity. All initial screening of L-albizziine provided evidence of its innocuous nature with *Locusta*; albizziine's potent insecticidal potential emerges indelibly.

These experiments were extended by Evans and Bell (1979) who used the same procedure to study *Anacridium melanorhodon*. This acridid is a feeding specialist of certain *Acacia* inhabiting Africa. For comparative purposes, *Locusta migratoria migratorioides* was also evaluated. *Anacridium melanorhodon* is considerably less sensitive to the tested amino acids, e.g., homoarginine and pipecolic acid, that occur in the leaves of its foodstuff (i.e., *Acacia*) than is the nonadapted *Locusta*. In general, the nonprotein amino acids stored in the seed are much more toxic to both test organisms than those proliferated and contained within the leaves. It is noteworthy that relatively high concentrations of these natural products, around 5%, are required for widespread expression of anti-feeding activity. By comparison, much lower levels of alkaloids are needed to create comparable toxicity (Janzen *et al.*, 1977).

Some phytotoxicity data of a comparative nature are also available. Wilson and Bell (1978, 1979) relied on germination as well as radicle and

---

*For most nonprotein amino acids, this corresponds to 1 to 2.5% (w/v).

Table 12   Comparative Toxicity of Some Naturally Occurring and Synthetic Nonprotein
Amino Acids[a,b]

Group I[c]
  L-Methionine-DL-sulfoximine
  L-Ethionine
  L-Azaserine
  L-3-(Methylenecyclopropyl)alanine
    (hypoglycin A)
  L-Allylglycine
  Se-methylseleno-DL-cysteine
  m-Fluoro-DL-tyrosine
  5-Fluoro-DL-tryptophan
  L-Canaline
  L-Canavanine
  O-Ureido-L-homoserine
  L-Azetidine-2-carboxylic acid
  Seleno-DL-ethionine
  Seleno-DL-methionine
  L-2-Amino-4-methylhex-4-enoic acid
  L-Djenkolic acid
  5-Methyl-DL-tryptophan

Group II[d]
  L-Glutamic acid hydrazide
  β-2-Thienyl-DL-alanine
  Seleno-DL-cystine
  5-Hydroxy-L-tryptophan
  O-Fluoro-DL-phenylalanine

Group III[e]
  p-Fluoro-DL-phenylalanine
  L-γ-Glutamyl-L-3-(methylenecyclo-
    propyl) alanine (hypoglycin B)
  Cycloserine
  3-Hydroxy-L-phenylalanine
  3,4-Dihydroxy-L-phenylalanine

  L-Canavaninosuccinic acid
  L-2,3-Diaminopropionic acid

Group IV[f]
  L-Norvaline
  DL-Azatryptophan
  L-Indospicine
  L-Albizziine
  L-2,4-Diaminobutyric acid
  N^6-Methyl-L-lysine
  O-Methyl-DL-serine

Group V[g]
  3-Cyano-L-alanine
  3-Fluoro-L-alanine
  3-Pyrazole-L-alanine
  4-Glutamyl-3-aminopropionitrile
  3-Methyl-DL-aspartic acid
  L-2-Amino-4-guanidinobutyric acid
  4-Hydroxy-4-methyl-L-glutamic acid
  2-Methyl-L-glutamic acid
  3-Methyl-L-glutamic acid
  4-Phenyl-L-glutamic acid
  L-Homoarginine
  L-Homocitrulline
  L-Mimosine
  2-Methyl-L-ornithine
  5-Methyl-L-proline
  4,5-Dimethyl-L-proline
  5-Phenyl-L-proline
  2-Methyl-DL-serine
  β-Phenyl-DL-serine

---

[a] Each amino acid was tested at a concentration of 1, 2, 4, 5, 10, 25, and 50 nmoles/ml with *L. minor* as previously described (Rosenthal *et al.*, 1975). Four fronds served as the inoculum; growth determinations were conducted after 6 and 10 days. The LD_{50} for each amino acid was calculated by log probit analysis (Finney, 1971). Reproduced with permission from the work of Gulati *et al.* (1981).

[b] The amino acids of groups I through IV are listed in order of decreasing toxicity; an alphabetical listing is employed for group V.

[c] LD_{50} less than 5 nmoles/ml.

[d] LD_{50} between 5 and 10 nmoles/ml.

[e] LD_{50} between 10 and 25 nmoles/ml.

[f] LD_{50} between 25 and 50 nmoles/ml.

[g] LD_{50} greater than 50 nmoles/ml.

hypocotyl growth of lettuce plants obtained from seeds that had hydrated in a solution of the appropriate compound. At a 1 mM concentration, only canavanine, albizziine, and N^β-oxalyl-α,β-diaminopropionic acid show an ability to curtail subsequent germination; successful seed germination plummets as the dose level increases to 10 mM. In the case of 2,4-diaminobutyric acid, increasing the dose from 1 mM as compared to 10 mM elicited a particularly strong enhancement in toxicity.

The nonprotein amino acids that caused more than 50% inhibition of radicle or hypocotyl growth when provided at 1 mM are 2-amino-4-methylaminopropionic acid, *2-amino-3-oxalylaminopropionic acid, azetidine-2-carboxylic acid, canavanine,* 2,3-diaminopropionic acid, 2,4-diaminobutyric acid, L-dopa, homoarginine, *mimosine, pipecolic acid,* and homoserine (the italicized compounds are the most potent) (Wilson and Bell, 1978). A follow-up study permitted the addition of the following natural products to this listing: 4-hydroxyarginine, enduracididine, 4-hydroxyproline, tetrahydrolathyrine, 5-hydroxynorleucine, and baikiain (Wilson and Bell, 1979).

An extensive evaluation of nonprotein amino acid phytotoxicity has also been conducted in the author's laboratory utilizing frond production by the aquatic microphyte, *Lemna minor* (Gulati *et al.,* 1981). Inspection of the LD_{50} values for these naturally occurring or synthetic compounds (Table 12) illustrates the marked toxicity of such methionine antagonists as methionine sulfoximine, and selenomethionine, the constituents of canavanine metabolism, certain fluorinated derivatives, hypoglycin, azetidine-2-carboxylic acid, and djenkolic acid. On the other hand, 2,4-diaminobutyric acid, indospicine, albizziine, mimosine, and β-cyanoalanine are conspicuous in their inability to elicit appreciable phytotoxicity in this aquatic plant.

G. CONCLUSIONS

1. Anomalous Protein Production

That some nonprotein amino acids are potent toxicants able to exert their debilitating effects in a host of organisms has been amply documented in the preceding sections. Of the many plausible rationales for their toxic action, anomalous protein production is proposed consistently to account for those substances bearing structural analogy to the constituents of proteins. Protein amino acids are activated by an ATP-dependent reaction mediated by a group of aminoacyl-tRNA synthetases.

$$\text{Amino acid + ATP + enzyme} \overset{Mg^{2+}}{\rightleftharpoons} \text{aminoacyl-AMP-enzyme + PP}_i$$

$$\text{Aminoacyl-AMP-enzyme + tRNA} \rightleftharpoons \text{aminoacyl-tRNA + enzyme + AMP}$$

The correct translation of the information contained within the genome requires a marked reaction specificity, for once aminoacylation is complete, the activated amino acid is incorporated into the polypeptide chain. Plants that produce nonprotein amino acids bearing structural similarity to protein constituents appear to exclude these particular natural products from their proteins, presumably due to the discriminatory capacity of their amino acid-activating enzymes. Thus, azetidine-2-carboxylic acid, the principal nitrogen-storing compound of *Convallaria majalis*, is not changed by the prolyl-tRNA synthetase of this species even though *C. majalis* stores 50 times the concentration demonstrated to be lethal to *Phaseolus* (Peterson and Fowden, 1965).

A more dramatic illustration of this point is provided by a study of Lea and Fowden (1972) in which glutamyl-tRNA synthetase was purified from *Phaseolus aureus, Hemerocallis fulva*, and *Caesalpinia bonduc*. *Phaseolus aureus* does not synthesize either the *erythro* or the *threo* forms of γ-methylated, γ-hydroxylated, or γ-methyl-γ-hydroxylated derivatives of glutamic acid, but it activates all these natural products. This enzyme, obtained from *Hemerocallis*, is able to activate the *erythro* diastereoisomer but not *threo*-γ-hydroxy-L-glutamic acid; in *Hemerocallis* only the *threo* compound is a natural product. On the other hand, *Caesalpinia bonduc* has a glutamyl-tRNA synthetase that aminoacylates the *threo*-L-isomer of γ-methylglutamic acid but not the naturally occurring *erythro*-γ-methyl derivative of glutamic acid (Table 13).

This ability to distinguish between diastereoisomers reflects the general ability of higher plants to avoid autotoxicity by not incorporating amino acid analogs into proteins (Fowden and Lea, 1979). It stands to reason that producer species would circumvent disrupted protein synthesis if they developed aminoacyl-tRNA synthetases possessing an active site that discriminates between a given protein amino acid and its structural analog. Experimental evidence suggestive of active site modification has been secured by studies of the prolyl-tRNA synthetase of azetidine-2-carboxylic acid-producing legumes. This enzyme cannot aminoacylate azetidine-2-carboxylic acid, but readily activates *exo*(*cis*)-3,4-methano-L-proline (Table 14). Nonproducing species esterify azetidine-2-carboxylic acid to the cognate tRNA of proline but exhibit little activity with *exo*(*cis*)-3,4-methanoproline, a sterically bulkier compound. These findings are consistent with a larger active site for the prolyl-tRNA synthetase of plants that have adapted to the production and storage of azetidine-2-carboxylic acid (Norris and Fowden, 1972).

Table 13 Kinetic Parameters Determined for Glutamic Acid and Several Analogs Using Glutamyl-tRNA Synthetase Preparations from Higher Plants[a]

Species	L-Glutamic acid		Erythro-γ-methyl-L-glutamic acid		Threo-γ-methyl-DL-glutamic acid		Threo-γ-hydroxy-L-glutamic acid		Erythro-γ-hydroxy-DL-glutamic acid		2(S), 4(S)-γ-Hydroxy-γ-methylglutamic acid	
	K_m	V_{max}	K_m	V_{max}	K_m	V_{max}	K_m	V_{max}	K_m	V_{max}	K_m	V_{max}
Phaseolus aureus seed	7.2×10^{-3}	100	1.55×10^{-2}	68.1	—	55.2 (75 mM)	2.11×10^{-2}	54.7	—	58.2 (75 mM)	3.43×10^{-2}	42.2
Hemerocallis fulva leaf	5.24×10^{-3}	100	2.81×10^{-2}	40.2	∞	0	∞	0	—	34.2 (75 mM)	1.25×10^{-1}	Calculated as 10.2
Caesalpinia bonduc seed	9.3×10^{-3}	100	∞	0	—	20.1 (75 mM)	5.21×10^{-2}	23.6	∞	0	∞	0

[a] The K_m values are expressed as molar concentrations with respect to the L form. The V_{max} values are expressed as percentages of the values determined for glutamic acid. Reproduced with permission from Fowden and Lea (1979).

Table 14 Kinetic Parameters of Prolyl-tRNA Synthetases from Azetidine-2-Carboxylic Acid Producer and Nonproducer Plants[a]

Species	Production of large amounts of azetidine-2-carboxylic acid	L-Proline, K_m ($\times 10^{-4}$ M)	L-Azetidine-2-carboxylic acid		cis-3,4-Methano-L-proline	
			K_m ($\times 10^{-3}$ M)	V_{max}[b]	K_m ($\times 10^{-3}$ M)	V_{max}[b]
Parkinsonia aculeata	Yes	4.35	∞	0–5	7.1	42
Delonix regia	Yes	1.82	∞	0–5	4.6	22
Convallaria majalis	Yes	4.5	∞	0–5	2.5	36
Beta vulgaris	No	4.5	2.2	73	ND[c]	<3
Hemerocallis fulva	No	6.25	5.3	75	ND[c]	<3
Phaseolus aureus	No	1.37	1.43	55	ND[c]	<2
Ranunculus bulbosa	No	2.9	2.0	66	∞	0

[a] All data derived from measurement of the ATP–PP_i exchange reaction as described by Norris and Fowden (1972). Reproduced with permission from Fowden and Lea (1979).

[b] The V_{max} values are expressed as a percentage of that obtained for proline.

[c] ND, Not determined.

Formation of aberrant, analog-containing protein undeniably represents the most frequently cited basis for the antimetabolic properties of certain toxic nonprotein amino acids. Disfunctional proteins are able potentially to disrupt virtually all aspects of plant growth, development, and reproduction. The real question, however, is how effectively does a canavanine- or azetidine-2-carboxylic acid-containing protein function? The difficulty in accurately assessing this question arises from the paucity of information on exactly what occurs when an analog is substituted for its protein amino acid counterpart. Isolation of alkaline phosphatase from *E. coli* in which 85% of the proline residues were replaced by 3,4-dehydroproline affected the heat lability and ultraviolet spectrum of the protein but the important criteria of catalytic function such as the K_m and V_{max} were unaltered (Fowden *et al.*, 1967). Massive replacement of methionine by selenomethionine in the β-galactosidase of *E. coli* also failed to effect the catalytic activity. Canavanine facilely replaces arginine in the alkaline phosphatase of *E. coli*; at least 13 and perhaps 18 or 19 of 20 to 22 arginyl residues are substituted. This massive replacement by canavanine caused subunit accumulation since they did not dimerize to yield the active enzyme (Attias *et al.*, 1969). Nevertheless, these workers stated: "There was also formed, however, a significant amount of enzymatically active protein in which most of the arginine residues had been replaced by canavanine." An earlier study in which either 7-azatryptophan or tryptazan replaced tryptophan resulted in active protein comparable to the native enzyme (Schlesinger, 1968).

Inclusion of nonprotein amino acids in the primary structure of a polypeptide need not manifest a dramatic or even significant affect on the functional properties of the macromolecule. Any residue not part of or contributing directly to the maintenance of the three-dimensional orientation of the active site would be expected to be replaced with some measure of impunity. That a hierarchy of importance for contributing amino acid residues exists is well known; certain replacements are simply more critical than others.

What emerges therefore is the realization that errant inclusion of customarily nonprotein amino acids into proteins can have a pervasively disruptive effect on overall protein function but the relevant experimental data permitting a quantitative assessment of this effect are still far too limited at this time. Studies of canavanine and protein production in insects bear meaningfully on this question. Tobacco hornworm, *Manduca sexta*, is very sensitive to the insecticidal effects of this arginine analog (Dahlman and Rosenthal, 1975). Newly ecdysed fifth stadium larvae assimilate radioactive canavanine into their newly synthesized protein, but the level of canavanine incorporated evidently is insufficient to cause preferential degradation of the resulting proteins. That is to say,

[³H]leucine-labeled proteins produced in the presence of canavanine turn over at the same rate as those formed in the absence of canavanine. There is abundant evidence that prokaryotic and eukaryotic organisms selectively degrade their aberrant proteins, e.g., those containing canavanine.

Automated amino acid analyses of the hydrolysate of highly purified arginine kinase, isolated from larvae administered appreciable canavanine, does not itself contain detectable canavanine (unpublished observations). These studies with *Manduca sexta* provide experimental evidence that under conditions where canavanine is toxic and detectable radioactive canavanine is incorporated into newly synthesized proteins, appreciable canavanyl protein formation may not occur.

A different picture emerges from studies of the migratory locust, *Locusta migratoria migratorioides*. The fat body of the female locust can be maintained in a chemically defined, sterile medium that continues to support protein biosynthesis and secretion. Most of the protein produced by the fat body is vitellogenin which is transported by the hemolymph to the oocyte where it contributes to vitellin production, the principal protein of the egg yolk. Vitellogenin produced in the presence of canavanine exhibits an electrophoretic mobility different from that of the native protein. Amino acid analysis of the protein hydrolysate reveals a distinctive canavanine peak and it appears that about 10% of the arginyl residues are replaced by canavanine (Pines *et al.,* 1981). The germane question of whether the inclusion of canavanine into this important insectan protein has adversely affected its biological properties has not been determined at this time. Thus, canavanine can but does not necessarily replace arginine in the protein of canavanine-treated organisms. It is not yet known to what extent inclusion of canavanine or some other nonprotein amino acid into the protein affects its normal catalytic, regulatory, or functional properties. In conclusion it is premature to assert that anomalous protein production is the *principal* basis for the antimetabolic properties of certain nonprotein amino acids that bear structural analogy to the constituents of proteins. Further experimental effort in this realm is required.

2. Additional Modes of Action

Nonprotein amino acids exhibit other common modes of action beside anomalous protein production. Some complex with the coenzyme to block essential catalytic action. Enzyme function is also inhibited competitively by virtue of the structural analogy of certain nonprotein amino acids to the natural substrate molecule; noncompetitive inhibition also occurs. Many compounds adversely affect critical reactions of macromolecular metabolism and alter various aspects of DNA, RNA, and/or protein synthesis and function. Impairment of normal protein and/or nucleic acid reactions strikes directly at fundamental processes of the living cell. This is demonstrated by their cytotoxic, hepatotoxic, and teratogenic properties. Nonprotein amino acid specificity of action is reflected by their ability to induce cirrhosis of the liver, cleft palate formation, and affect essential reactions of collagen formation. These compounds can disrupt the estrous cycle, diminish overall reproduction, and exhibit general abortifacient effects. They can have a devastating influence on insect development processes leading to massive cellular deformity and a significantly lower incident of larval–pupal metamorphosis and cause severe growth inhibition in a host of higher plants, bacteria and other prokaryotes, and fungi.

Among the other common modes of action are disruption of amino acid uptake and translocation, generation of erroneous repression signals, false end-product inhibition, curtailment of organelle formation and function, alteration in cellular structural components, and impairment of energy-producing reactions (for additional information on this point, see Lea and Norris, 1976; Fowden and Lea, 1979; Rosenthal, 1977a).

With so many concurrent processes potentially affected, it is impractical, if not impossible, to weigh the consequences of each factor isolated from the others. Doubtless, these factors interact and reinforce each other to create the totality of the toxicosis which is collectively referred to as the compound's antimetabolic properties. As such, nonprotein amino acid toxicity must be viewed as a multitude of effects on many levels of organization in the living organism.

3. Plant-Herbivore Interaction

The diversity, intensity, and efficacy of the biological manifestations of toxic nonprotein amino acid consumption lends itself to the speculation that these compounds afford to the producing plant a significant measure of protection from herbivory, pathogenicity, and competition from other plants. While direct evidence linking toxic secondary plant metabolites with enhanced biological fitness is admittedly limited, a considerable body of inferential data has been amassed (see reviews by Rhoades, 1979; McKey, 1979). Concern over the consequences of eating noxious plant products is directed primarily at ourselves or animals that participate in human economic endeavors. Although understandable, it has nevertheless created a biased picture of the actual range of effectiveness of toxic metabolites in living systems. A growing consensus has emerged that the evolutionary development of deleterious compounds is more correctly viewed as reflecting long-term contact between plants and such organisms as insects, small rodents, nematodes, and molluscan herbivores as well as various pathogenic bacteria and fungi. Of the many adversaries that have contributed to present day control and containment strategies predicated on the secondary chemistry of the plant, insects are, and presumably were, the most aggressive, tenacious, and destructive combatants. The pervasive influence of insects and plants on each other, to a point that they are taken to have coevolved, has been stated most elegantly and lucidly by Paul Feeny (1975) in the context of his consideration of the classical study by Ehrlich and Raven (1965) on the development of feeding patterns:

> According to this hypothesis, now generally accepted, at least some of the secondary substances found in plants were evolved or elaborated in response to attack by insects. Some of the associated insects evolved methods of tolerating the new plant chemicals and were thus able to remain associated with their particular host species. As plants evolved through time, some of their associated insects coevolved with them, often leaving relatives on the original plant families or genera. One man's meat became another man's poison and present day insects are adapted to tolerate only a certain range of chemicals and therefore only a certain range of plants. Evolving along with detoxication mechanisms, chemosensory systems responding differentially to secondary compounds enabled insects to locate the plants to which they were chemically adapted. We are thus witnessing an evolutionary arms race in which the plants, for survival, must deploy a fraction of their metabolic budgets on defense (physical as well as chemical) and the insects must devote a portion of their assimilated energy and nutrients on various devices for host location and attack.

In essence, the feeding activity of phytophagous insects and other herbivores and the ravages of a host of pathogenic organisms probably provided the selective pressure for plant elaboration and maintenance of sophisticated chemical barriers to predation and disease. The possibility that organisms presently unknown or unappreciated contributed significantly to the formation of these chemical barriers must not be overlooked. Natural selection capitalized on the inherent variability in secondary plant chemistry to favor those capable of mounting an effective chemical defense that would enhance their intrinsic biological fitness. Early defensive modes stimulated herbivore adaptation and ultimately instigated appropriate strategies of counteradaptation. Whittaker and Feeny (1971) have aptly characterized this perpetual struggle as a continuous investment of the metabolic resources of the plant against sustained countermeasures by herbivores. Nothing else seems so reasonable in rationalizing plant survival in the face of aggressive animal herbivory, ubiquitous pathogenic organisms, and other competitive elements.

In espousing the soundness of this concept of a defensive role for certain natural products, I am guided by the fact that it is untenable for me to view secondary plant metabolites as plant "garbage cans" designed for "containerizing" nonfunctional molecules. Quite the contrary, Seigler and Price (1976) and Seigler (1977) have made the germane point that many secondary metabolites actually have primary metabolic functions; they turn over rapidly and shunt carbon and nitrogen into metabolically active pools. Many nonprotein amino acids, including toxic ones, sustain other metabolic pools. They are not merely static components of the plant's metabolism assembled conveniently to store unwanted atoms in an isolated cellular recess, but a dynamic part of overall plant carbon and nitrogen metabolisms. As part of that dynamic role, many also function in the chemical defense of certain higher plants.

The specific bases for the assertion that toxic nonprotein amino acids contribute to the secondary plant defensive compound complement have recently been stated by Bell (1978) and Rosenthal and Bell (1979). In developing these explicit statements, we drew on the fundamental ecological studies of many workers including Janzen (1969), Freeland and Janzen (1974), Rehr et al. (1973a,b), and others. In essence, the contention that nonprotein amino acids play protective roles rested on the following lines of evidence. First, there is the intrinsic toxicity exhibited by these secondary metabolites in a wide range of herbivorous and pathogenic organisms. Second, the very high concentration of these natural products that accumulate within the plant represents an expenditure of metabolic resources. It is difficult to rationalize the synthesis and storage of massive quantities of nitrogen-rich compounds without some commensurate benefit in overall plant fitness. Except for a limited group of plants that include only certain legumes, the vast majority of higher plants lack the symbiotic relationship essential to diatomic nitrogen fixation into ammonia. Rarely overabundant, nitrogen is often rate limiting to plant growth, hardly a nutrient to be wasted capriciously. That carbon skeleton synthesis for amino acid formation is an energy utilizing and storing process is revealed by the ability of amino acids to sustain respiratory reactions when customary caloric reserves are denied. Unless these toxic molecules exhibit a facile movement into other metabolic pools so that they can be amalgamated with the nitrogen-storing and translocating metabolites, they must be viewed as an investment against increased predation of photosynthetically active tissues, loss of the vital seed crop, or diminished competition for space, water, and nutrients. Or some such appropriate benefit for the plant that will improve its Darwinian fitness.

An intriguing corollary to this point develops in an attempt to understand the basis for large-scale investment of plant nitrogen for chemical

Table 15. Total Nitrogen and Amino Acid Nitrogen Committed to Canavanine Storage in the Seed of Certain Legumes[a]

Species	Seed canavanine content (% dry wt.)	Seed nitrogen content (% dry wt.)	Nitrogen stored in canavanine (%)	Amino acid nitrogen stored in canavanine (%)
Colutea arborescens	5.72	8.79	20.7	82
Caragana arborescens	6.06	6.69	28.8	83
Vicia gigantea	6.63	6.32	33.4	86
Canavalia maritima	6.25	5.14	38.7	80
Robinia pseudocacia	9.75	7.37	42.1	91
Dioclea megacarpa	12.71	7.41	54.6	94
Wisteria floribunda	12.26	7.01	55.6	96

[a] From Rosenthal (1977c).

defense of certain seeds. The data of Table 15 reveal the total nitrogen and soluble amino acid nitrogen of certain legumes committed to the synthesis and eventual storage of canavanine in the seed. One can surmise that mutation, genetic recombination, or some other chance event contributed to the formation of the genes enabling canavanine's synthesis. Perhaps, this was achieved by subtle alteration in the genetic complement responsible for producing arginine. Selection pressure of herbivores and pathogens may have escalated the amount of plant nitrogen diverted into seed canavanine production (Rosenthal, 1977c). With canavanine's marked toxicity, this tactic may have achieved a protective effect with such success that alternative modes of secondary plant defense were compromised severely or even abandoned.

This scenario gains added credence, particularly in the case of canavanine, because its nitrogen does move easily into other metabolic pools. It is cleaved enzymatically by arginase to canaline and urea. The latter product is hydrolyzed by urease to carbon dioxide and ammonia; the released nitrogen can then be stored as amide amino acid nitrogen (Rosenthal, 1970). A correlation exists between the level of stored seed canavanine and urease content (Rosenthal, 1974). In other words, the storage of 13% or more of canavanine not only represents a formidable chemical barrier to predation, but also a readily tapped source of stored nitrogen for the growing seed. Canavanine possesses all of the metabolic flexibility of arginine while providing the concurrent benefit of appreciable antiherbivore activity.

This point also serves as a preamble to the much more general matter of toxic nonprotein amino acids that are structural analogs of their protein-containing amino acid counterparts. Since the genes directing the synthesis and utilization of the latter compound already exist, it is difficult to interpret the massive accumulation of so many nonprotein amino acids uniquely in terms of nitrogen storage when their related protein amino acids perform this function just as effectively. I simply do not believe that storage and protection should be viewed as mutually exclusive functions.

Third, it is almost always found that nonprotein amino acids are manufactured and stored in very ample quantities in the vegetative tissues; the seed is not the sole depository for these natural products. Plant nitrogen resources would be more effectively conserved if these compounds were produced at the time of flower anthesis and development. This resource investment is rendered more explicable in terms of nonprotein amino acid capacity to provide a chemical barrier against predation in the vegetative tissues.

Fourth, the fact that the concentration of certain nonprotein amino acids within the seed population of a given species is often fixed creates the possibility that it may be the result of two opposing influences. The upper concentration limit results from the competition between plants that favors those members with a nitrogen resource allocation adequate to achieve the necessary protective effects. The lower level can reflect predator pressure that eliminates individuals from the population with less than the requisite nonprotein amino acid content required to deter predation adequately (Janzen, 1969).

A fifth line of evidence is provided by the incisive studies of Janzen (1969) demonstrating that the seed crop weight per unit area of canopy is inversely proportional to the seed size. Chemical examination of the appropriate seeds disclosed that the larger seeds contain these secondary metabolites while the smaller individuals do not. These smaller-seeded species are subject generally to bruchid beetle attack while the larger-seeded samples enjoy much greater freedom from predation. This finding suggests that two strategies are operative in curtailing seed predator activity. The first is production of a copious number of smaller seeds, thereby ensuring that a portion of the total crop will escape predation. The second relies on production and accumulation of insecticidal seed amino acids. In these instances, the metabolic cost to the plant for producing these compounds is probably less than the expense of generating a larger seed crop to compensate for predator attrition.

Janzen (1969) also reported that seed-eating beetles which forage for seeds that store a particular nonprotein amino acid do not survive when reared on seeds containing other members of this group of secondary metabolites. This is an important finding since it suggests a marked degree of specificity in insect adaptation to a particular non-protein amino acid. A general acquired resistance to these secondary plant metabolites apparently does not occur and these compounds continue to afford protection against the vast majority of potential predators.

Finally, if the above contentions are valid, then it should be possible to identify herbivores that react to the chemical barriers represented by toxic nonprotein amino acids by either developing effective mechanisms of biochemical detoxification or in some other way counteracting these metabolic obstacles. This has been achieved in a series of fundamental studies of the bruchid beetle *Caryedes brasiliensis* and the legume *Dioclea megacarpa* described in detail elsewhere (see p. 102).

I have presented a consideration of the factors that have served as the basis for a growing belief that nonprotein amino acids provided an adaptive advantage to the plant in the continued struggle for survival against various predators. I have not forgotten that there are few actual field data supporting the ability of these natural products to reduce herbivore feeding activity, minimize microbial activity, or enhance overall plant fitness. That there are many lacunae in our knowledge of insect interaction at the level of secondary plant metabolites is only too evident. Nevertheless, I selected to emphasize the rising tide of suggestive findings that are wholly consistent with the view that toxic nonprotein amino acids contribute significantly to the chemical defense of higher plants.

Before closing this chapter, it is evident that throughout the consideration of the toxic nonprotein amino and imino acids, information was provided on the known distributional pattern for these compounds. Statements on the biological sources of these natural products must be viewed in the light of an interesting paper by Leslie Fowden (1972) which is noteworthy for its bearing on secondary metabolite occurrence and its employment in establishing phylogenetic relationships. As part of the industrial purification of sucrose from sugar beet, the nitrogenous constituents were removed and collected by ion-exchange chromatography. This industrial process permitted analysis of the nonprotein amino acid complement from an enormous amount of plant material (10^9 kg). It revealed the presence of the N^γ-acetyl and N^γ-lactyl amino acid derivatives of α,γ-diaminobutyric acid. The former was known only from *Euphorbia pulcherrima* while the latter represents the only known naturally occurring lactyl derivative. In addition, N^ϵ-acetyllysine was established as a new natural product as was N^ϵ-acetyl-*allo*-δ-hydroxylysine. Azetidine-2-carboxylic acid was also shown to be a constituent of sugar beet, a member of the Chenopodiaceae—phylogenetically removed from the Liliaceae, the commonly recognized natural sources. The γ-glutamyl derivative of γ-aminobutyric acid was also isolated; it had been known from only one previous source. Thus, this higher plant extract possessed three novel nonprotein amino acids and three others believed previously not to be synthesized by members of this family of higher plants.

Responding to these disclosures, Fowden stated: "The isolation of these compounds was possible only because extremely large quantities of plant material were initially processed in an industrial plant. As a consequence of this approach, a number of compounds are now recognized as constituents of the sugar beet plant, although it is certain that they would have remained uncharacterized if only conventional laboratory procedures had been employed." Continuing further, Fowden explained that plant inability to accumulate detectable levels of a specific compound need not reflect an absence of the requisite enzymes for their production. After considering these assertions, Janzen (1979) contended that the occurrence of minute levels of a particular metabolite from an enormous starting material can, in fact, represent an artifact of isolation or limited production due to a lack of absolute substrate specificity or degradation or some non-enzymatic process. Janzen is justified in coercing us to be mindful that we may be "skating on thin ice" since one cannot deny the relative paucity of the isolated compounds nor the basic soundness of some of his arguments. While his admonitions have substance and serve a useful function, there is obvious merit to Fowden's basic premise that the absence of a particular metabolite from a given higher plant sample cannot be taken as evidence in delineating phylogenetic relationships and that we must be ever mindful of how we are examining and analyzing natural product occurrence as well as what it is we find.

ADDITIONAL READING

Bell, E. A. (1972). Toxic amino acids in the Leguminosae. *In* "Phytochemical Ecology" (J. B. Harborne, ed.), pp. 163–177. Academic Press, New York.

Bell, E. A. (1973). Amino acids of natural origin. *In* "Amino Acids, Peptides, and Related Compounds" (D. H. Hey and D. I. Johns, eds.), Vol. 6, pp. 1–16. Butterworth, London.

Bell, E. A. (1976). 'Uncommon' amino acids in plants. *FEBS Lett.* **64,** 29–35.

Bell, E. A. (1977). The possible significance of uncommon amino acids in plant-vertebrate, plant-insect, and plant-plant relations. *Pontif. Acad. Sci. Scr. Varia* **41,** 571–595.

Bell, E. A. (1980). Non-protein amino acids in plants. *In* "Secondary Plant Products" (E. A. Bell and B. V. Charlwood, eds.), Encycl. Plant Physiol., New Ser., Vol. 8, pp. 403–432. Springer-Verlag, Berlin and New York.

Bell, E. A. (1980). The non-protein amino acids of higher plants. *Endeavour* **4,** 102–107.

Bell, E. A. (1980). The non-protein amino acids occurring in plants. *Prog. Phytochem.* **7,** 171–196,

Fowden, L. (1970). The non-protein amino acids of plants. *Prog. Phytochem.* **2,** 203–265.

Fowden, L. (1973). Amino acids. *Phytochemistry* **2,** 1–29.

Fowden, L. (1974). Non-protein amino acids from plants: Distribution, biosynthesis, and analog functions. *Recent Adv. Phytochem.* **8,** 95–122.

Fowden, L. (1976). Amino acids: Occurrence, biosynthesis and analogue behavior in plants. *Perspect. Exp. Biol.* **2,** 263–272.

Fowden, L., Lewis, D., and Tristram, H. (1967). Toxic amino acids: Their action as antimetabolites. *Adv. Enzymol.* **29,** 90–163.

Fowden, L., Lea, P. J., and Bell, E. A. (1979). The non-protein amino acids of plants. *Adv. Enzymol.* **50,** 117–175.

Hegarty, M. P., and Peterson, P. J. (1973). Free amino acids, bound amino acids, amines and ureides. *In* "Chemistry and Biochemistry of Herbage" (G. W. Butler and R. W. Bailey, eds.), Vol. 1, pp. 1–62. Academic Press, New York.

Hylin, J. W. (1969). Toxic peptides and amino acids in foods and feeds. *J. Agric. Food Chem.* **17,** 492–496.

Kjaer, A., and Larsen, P. O. (1973). Non-protein amino acids, cyanogenic glycosides, and glucosinolates. *Biosynthesis* **2,** 71–105.

Kjaer, A., and Larsen, P. O. (1976). Non-protein amino acids, cyanogenic glycosides, and glucosinolates. *Biosynthesis* **4,** 179–203.

Kjaer, A., and Larsen, P. O. (1977). Non-protein amino acids, cyanogenic glycosides, and glucosinolates. *Biosynthesis* **5,** 120–135.

Kjaer, A., and Larsen, P. O. (1981). Non-protein amino acids, cyanogenic principles, and glucosinolates. *Biosynthesis* **8,** 104–121.

Lea, P. J. (1978). Biosynthesis of unusual amino acids. *Int. Rev. Biochem.* **18,** 1–47.

Lea, P. J., and Norris, R. D. (1976). The use of amino acid analogues in studies of plant metabolism. *Phytochemistry* **15,** 585–595.

Murti, V. V. S., and Seshadri, T. R. (1964). Toxic amino acids of plants. *Curr. Sci.* **11,** 323–329.

Norris, R. D., and Lea, P. J. (1976). The use of amino acid analogues in biological studies. *Sci. Prog. Oxf.* **63,** 65–85.

Rosenthal, G. A., and Bell, E. A. (1979). Naturally occurring, toxic non-protein amino acids. *In* "Herbivores—Their Interaction with Secondary Plant Metabolites" (G. A. Rosenthal and D. H. Janzen, eds), pp. 353–385. Academic Press, New York.

Thompson, J. F., Morris, C. J., and Smith, I. K. (1969). New naturally occurring amino acids. *Annu. Rev. Biochem.* **38,** 137–158.

Chapter 4

Components of Intermediary Metabolism

Compounds (amino and imino acids) present in low concentration (often below the threshold levels of detection) are frequently overlooked and therefore rarely isolated and characterized. There are then probably hundreds of additional types of amino acids, elaborated by plants, that await identification; and it is equally probable that many, if not all, of the compounds so far characterized have a much wider distribution within the plant kingdom, albeit in lower concentration than is realized at present.

Leslie Fowden, 1973

A. NITROGEN ASSIMILATION

Consideration of the details of the intermediary metabolism of non-protein amino acids would be incomplete without first developing the reactions whereby diatomic nitrogen, fixed as ammonia, is assimilated into amino acid pools sustaining these reactions. Our present appreciation of these factors results to a large extent from the researches and writings of Miflin and Lea (1977). The classical concept of ammonia assimilation centered on glutamic acid dehydrogenase-mediated reductive amination of 2-oxoglutaric acid to glutamic acid. This reaction funnels carbon skeleton generated by the Krebs cycle into nitrogen metabolism but it does so with a relatively poor affinity for ammonia. It is now appreciated that amination of 2-oxoglutaric acid by glutamine in conjunction with glutamine synthetase represents a viable alternative to the glutamic acid dehydrogenase-catalyzed reaction for assimilating ammonia.

$$\text{L-Glutamine} + \text{2-oxoglutaric acid} \rightarrow 2 \text{ L-glutamic acid}$$

The enzyme fostering this assimilatory route utilizes a reducing agent such as NADH, NAD(P)H, or reduced ferredoxin and is known trivially as glutamate synthase but the acronym GOGAT (glutamine:2-oxoglutarate aminotransferase) has come to dominate the current literature (see Dougall, 1974; Fowler *et al.*, 1974; Lea and Miflin, 1974).

The details of glutamine production from glutamic acid, particularly the stereochemical aspects, have been developed in a series of outstanding investigations by Meister and associates (see Tate and Meister, 1973). In essence, glutamic acid is aminated after an ATP-dependent activation:

$$\overset{\text{Mg}^{2+},\ \text{Mn}^{2+}}{\text{L-glutamic acid} + \text{ATP} + \text{NH}_3 \rightleftharpoons \text{L-glutamine} + \text{ADP} + \text{P}_i}$$

Once ammonia forms the amide group of glutamine, this amino acid becomes a principal nitrogen donor for the plant's biosynthetic reactions. The activity of GOGAT in association with glutamine synthetase (GS) provides a means for effectively sequestering NH_3 as glutamine in an irreversible reaction. Indeed, it is appropriate to view these ammonia assimilating reactions as being cyclic in nature:

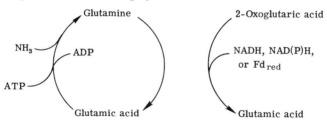

Adapted from Miflin and Lea (1980).

Additionally, glutamic acid requirements are most efficiently met by the GOGAT system since the ammonia K_m for glutamine synthetase (1–$2 \times 10^{-5}M$) is much lower than it is with glutamic acid dehydrogenase (GDH) (5–$70 \times 10^{-3}M$). This factor is significant under conditions of limited ammonia availability.

Studies of the above enzymes obtained from the root of *Zea mays* have resulted in a somewhat modified picture of nitrogen assimilation (Oaks *et al.*, 1980). GS and GOGAT were found to be the principal nitrogen assimilatory enzymes in the young apical region of intact root while GDH and asparagine synthetase were more active in the older, more matured regions of the root. A marked enhancement of GDH and asparagine synthetase was reported when excised root tips were exposed to NH_4^+ and NO_3^-; at the same time, GOGAT and GS activities were reduced somewhat. Stewart and Rhoades (1977) and Rhoades et al. (1976) have provided evidence from Lemna minor that GS and GOGAT activities predominate when NH_4^+ availability is low while GDH is enhanced as the NH_4^+ concentration increases. ATP was shown to inhibit GDH while ADP and 5'-AMP curtailed GS but not GDH activity. The enzymes GS and GOGAT were taken, therefore, to be the principal components of ammonia assimilation when the cellular energy charge was high and ammonia limiting. Conversely, under reduced cellular energy charge, which would minimize assimilatory reactions, and elevated ammonia, GDH is apparently of greater importance in nitrogen assimilation.

Higher plant production of asparagine involves the enzyme-asparagine synthetase:

$$\text{L-aspartic acid} + \text{L-glutamine} + \text{ATP} \overset{Mg^{2+}}{\rightleftharpoons} \text{L-asparagine} + \text{L-glutamic acid} + \text{AMP} + PP_i$$

This reaction has been established as the principal basis for higher plant production of asparagine but an enzyme has been isolated from the roots of Zea mays that mediates active asparagine production with ammonia as the nitrogen donor (Stuben et al., 1979). These workers reported a K_m for glutamine of 1.0 mM and 2 to 3 mM for NH_4^+.

Mobilization of the nitrogen incorporated into asparagine is achieved by transamination to 2-oxosuccinamic acid. This reaction has been shown in soya bean (Streeter, 1977) and pea leaves (Lloyd and Joy, 1978). Hydrolysis of asparagine to yield aspartic acid is another important means for the mobilization of stored amide nitrogen. This reaction is catalyzed by asparaginase which has been purified from Lupinus polyphyllus (Lea et al., 1979). In a recent publication by Sodek et al. (1980) the workers at Rothamsted point out that asparaginase has not been demonstrated from several legumes analyzed for this important enzyme. They provide evidence that this apparent lack of enzyme production actually reflects an unrealized potassium dependence for the enzyme. In Lupinus albus, the asparaginase activity of the cotyledons increases from 0.08 μmoles/hr/gm fresh weight to 4.70 (nearly 60-fold) upon the addition of K^+ to the enzyme reaction mixture. In a similar way, Rognes (1980) has reported that such monovalent anions as Cl^- and Br^- strongly activate the glutamine-dependent asparagine synthetase of Lupinus lutens seedlings. There is no experimental evidence that asparagine can function directly as a nitrogen donor for the GOGAT reaction.

Some controversy exists over the ability of asparagine to supply nitrogen to the developing plant. In maturing *Lupinus albus* seed, 55–60% of the transported nitrogen is carried as asparagine and no more than 15% of this nitrogen is deployed for protein production involving asparagine. ^{14}C- and ^{15}N-amide labeling of asparagine and its administration at the time of peak seed protein production reveals that at least 60% of the nitrogen is transferred to a variety of amino acids. As may be expected, the carbon skeleton is far more stable and two-thirds of the ^{14}C remains with aspartic acid. Asparagine-derived nitrogen strongly supports amino acid synthesis for protein production and asparagine derived from the turnover of plant proteins is an important means whereby seed storage proteins maintain the nitrogen metabolism of the developing plant (see Pate *et al.*, 1965).

Studies of asparagine formation in soya bean nodules provided evidence that exogenous $^{15}NH_4^+$ was incorporated into the amide nitrogen of glutamine and then to the amide group of asparagine. Feeding detached nodules [^{15}N](amide)-glutamine resulted in the movement of the ^{15}N atom almost entirely into the amide nitrogen of asparagine (Fujihara and Yamaguchi, 1980). Ample evidence exists to support the assertion that in most legumes asparagine is the principal recipient of the nitrogen obtained by symbiotic nitrogen fixation (see Fujihara and Yamaguchi, 1980, for additional references).

B. THE ASPARTIC ACID FAMILY

Aspartic acid contributes to the production of numerous amino acids; for example, homoserine formation, and as a consequence: threonine, isoleucine, and methionine. β-Aspartylsemialdehyde radiates into lysine metabolism and this in turn contributes to amino acid reactions through lysine's precursor role in pipecolic acid formation and its probable contribution in forming hydroxylated derivatives of pipecolic acid. These reactions will be detailed in this section but several other compounds must also be mentioned, namely, the N^4-ethyl [16]*, N^4-methyl [17], and N^4-2-hydroxyethyl [43] derivatives of asparagine. The first is a lower homolog of theanine (N^5-ethylglutamine [25]). Theanine is produced in an ATP-dependent reaction whose mode of action parallels that of glutamine synthetase; glutamic acid and ethylamine are the substrates. In contrast, [U-^{14}C]aspartic acid is not converted to radioactive N^4-ethyl or N^4-hydroxyethyl derivatives of asparagine by *Ecballium* seedlings (Frisch *et al.*, 1967). The N^4-substituted asparagines appear to form by a transferase reaction involving asparagine and ethylamine or ethanol-

---

*The numeral discloses the compound's location in the Appendix.

amine rather than an ATP-dependent synthetase-type reaction. The hydroxylated derivatives of aspartic acid is limited to the *erythro* form of 3-hydroxyaspartic acid [44]. Thus, there are far fewer derivatives of aspartic acid and its amide than exist for the glutamic acid family.

1. Homoserine and Related Sulfur Metabolism

Homoserine enjoys essentially universal distribution by virtue of its key role in the biosynthesis of threonine, isoleucine, and methionine. While these reaction pathways are best known from microorganisms, they also occur in higher plants. Homoserine production from aspartic acid is initiated by a phosphorylation to yield 3-aspartylphosphate and reduction to 3-aspartylsemialdehyde. The latter compound need only be further reduced to generate homoserine. Each of the enzymes involved in these reactions has been isolated from a higher plant.

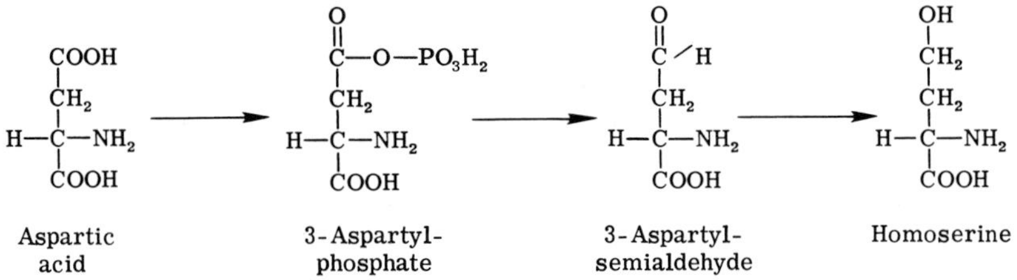

| Aspartic acid | 3-Aspartyl-phosphate | 3-Aspartyl-semialdehyde | Homoserine |

Homoserine is not a constituent of the ungerminated pea (*Pisum sativum*) seed but it increases dramatically during the first week or so of growth until it constitutes 70% of the soluble nitrogen and 12% of the dry seedling weight (Mitchell and Bidwell, 1970). In this legume, homoserine is as effective as asparagine or glutamine in conveying carbon and nitrogen to the growing regions of the plant. In fact, aspartic acid conversion to homoserine is completed in the root prior to translocation to the epicotyl.

Homoserine participation in methionine production, which represents an important bridge between amino acid carbon and sulfur metabolism, can occur by several distinctive pathways. Prior to considering this, however, it would be best to examine briefly the formation of cysteine which directly supports methionine production in higher plants.

Inorganic sulfur is reductively assimilated into higher plants and converted either to free or bound sulfide; the precise chemical species that is the immediate precursor of cysteine remains unknown. Many higher plants have been established as sources of L-serine acetyltransferase which mediates the formation of O-acetyl-L-serine from L-serine and acetyl-CoA (Ngo and Shargool, 1974). Studies of spinach extracts revealed a mechanism for higher plant biosynthesis of cysteine involving a direct sulfhydration reaction (Giovanelli and Mudd, 1967):

$$O\text{-Acetyl-L-serine} + \text{sulfide} \rightarrow \text{L-cysteine} + \text{acetate} \quad \text{(a)}$$

O-Acetyl-L-serine sulfhydrase (cysteine synthase), mediating reaction (a), has been purified to apparent homogeneity from rape, *Brassica chinensis,* and other plant sources including wheat seeds, *Triticum aestivum* (Ascaño and Nicholas, 1977). The *Brassica* enzyme possesses a mass of 62,000 daltons which is obtained from two identical subunits each containing a mole of pyridoxal phosphate (Masada *et al.,* 1975). Assays of the cysteine synthase of *Trifolium repens* and *Pisum sativum* indicated that 68 to 86% of the enzyme activity was associated with the chloroplastic stroma (Ng and Anderson, 1978). This is consistent with higher plant localization of amino acid metabolic reactions within the chloroplast. Interestingly, a lyase-type activity also occurs in *Brassica* that degrades O-acetylserine to pyruvic acid, acetate, and ammonia (Mazelis and Fowden, 1972). It is now well accepted that higher plant serine acetyltransferase and cysteine synthase are responsible for converting assimilated sulfate to cysteine via O-acetylserine. A serine sulfhydrase exists that can react serine directly with sulfide to form cysteine and water but this is not considered to be very important physiologically (Giovanelli *et al.,* 1980).

Higher plant production of methionine involves the reaction of cysteine with an α-aminobutyryl donor to form cystathionine. The latter serves as a precursor for methionine via homocysteine formation; there is little evidence of significant cysteine formation from cystathionine (Fig. 35). Our appreciation of the identity of the physiological α-aminobutyryl donor and many of the details of higher plant sulfur amino acid metabolism can be traced directly to the incisive research of John Giovanelli and associates at the National Institutes of Health.

Bacteria such as *Bacillus subtilis* and the fungus *Neurospora crassa* first condense homoserine with acetyl-CoA to form O-acetylhomoserine prior to cystathionine production (Fig. 35). This acylhomoserine derivative also accumulates in *Pisum sativum. Lathyrus sativus* is also distinctive in its storage of another O-acylhomoserine derivative, namely, O-oxalylhomoserine. However, these natural products are limited to the developing fruit (Giovanelli *et al.,* 1974), and this restricted distribution speaks against their functioning significantly in cystathionine production in these legumes. Similarly, succinylhomoserine synthesis has been reported in pea (Clandinin and Cossins, 1974), but this report has been challenged (Giovanelli *et al.,* 1980).

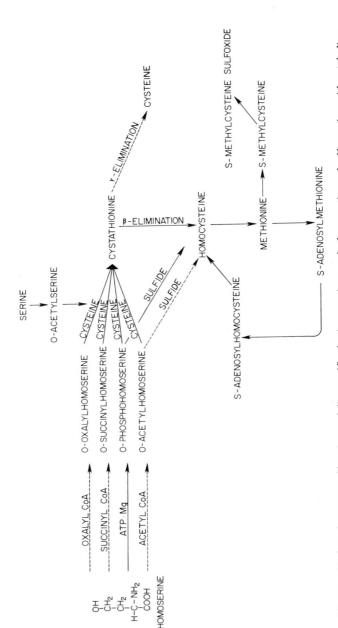

Fig. 35. Higher plant transsulfuration and direct sulfhydration reactions and other reactions of sulfur amino acid metabolism. Based primarily on the research of Giovanelli *et al.* (1974). The dashed arrows represent reactions that are not of principal importance in green plant production of cystathionine or homocysteine but which have been reported in certain plants.

Homoserine kinase mediates the phosphorylation of homoserine to *O*-phosphohomoserine:

$$\text{L-Homoserine} + \text{ATP} \rightarrow O\text{-phospho-L-homoserine} + \text{ADP} \quad \text{(b)}$$

Higher plant ability to conduct this reaction appears to be very widespread. This phosphorylated derivative is noteworthy in that only green plants can utilize it for cystathionine and homocysteine production. In these organisms, it is the physiological precursor of cystathionine (Datko *et al.*, 1974; see also Dougall and Fulton, 1967). Cystathionine can also form from *O*-succinylhomoserine but this reaction appears to be of importance primarily in certain bacteria such as *Salmonella typhimurium* and *E. coli* (Giovanelli *et al.*, 1974). In other words, while crude higher plant extracts can use such acyl derivatives as *O*-malonyl, *O*-oxalyl, *O*-succinyl, and even *O*-acetyl forms of homoserine for *in vitro* cystathionine formation, only the *O*-phosphoryl ester is of physiological significance (Giovanelli *et al.*, 1974).

In 1967, Giovanelli and Mudd determined that spinach could foster the production of methionine by direct sulfhydration:

$$O\text{-Acetyl-L-homoserine} + \text{sulfide} \rightarrow \text{L-homocysteine} + \text{acetate} \quad \text{(c)}$$

This work was significant for providing the first indication that *de novo* synthesis of homocysteine in plants could be achieved by direct sulfhydration. In contrast, *Neurospora* forms homocysteine from homoserine (Wiebers and Garner, 1963):

$$\text{Homoserine} + \text{sulfide} \rightarrow \text{homocysteine} + H_2O$$

By 1977, it was appreciated that in addition to reactions (a) and (c), green plants also mediated an *O*-phosphohomoserine-dependent sulfhydrase reaction. Once again, several acyl homoserine derivatives were active *in vitro* but inspection of various green plants, selected to represent the major phylogenetic divisions of the plant kingdom, established the physiological role of this phosphorylated derivative (Datko *et al.*, 1977). Thus, *O*-phosphohomoserine emerges as the principal physiological precursor of homocysteine formed by direct sulfhydration in green plants.

Higher plants are able to produce homocysteine by an alternative mechanism: *transsulfuration*. In this process, cysteine functions as the acceptor molecule in the transfer of an α-aminobutyryl moiety to yield cystathionine. The latter compound is subsequently cleaved to homocysteine (Fig. 35). As mentioned earlier, several acyl homoserine derivatives are active in the *in vitro* production of cystathionine but only *O*-phosphohomoserine is the physiological precursor of homocysteine and methionine production in higher plants (Giovanelli and Mudd, 1971; Datko et al., 1974).

Utilizing *Chlorella sorokiniana*, Giovanelli et al. (1978) compared the relative contribution of *transulfuration* and *direct sulfhydration* in the assimilation of $^{35}SO_4^{2-}$ into sulfur-containing amino acids. The resulting labeling pattern was consistent with the movement from cysteine to cystathionine to homocysteine. One second after ^{35}S administration, the ratio of [^{35}S]cysteine to [^{35}S]homocysteine was 127 but after 10 sec it approximated the isotopic equilibrium value of 4. This pattern is anticipated if cysteine and cystathionine function as homocysteine precursors. If cysteine and homocysteine result from direct sulfhydration from a commonly accessible pool of S^{2-}, the above ratio would have been less than unity prior to approaching the isotopic equilibrium value. Finally, the labeling of cystathionine itself reflected its predicted property of a rapidly turning-over intermediate. A maximum ceiling of 3% was placed on the possible contribution of direct sulfhydration to homocysteine biosynthesis (Giovanelli et al., 1978). The finding of these studies with *Chlorella* and additional studies with *Lemna* led Giovanelli and associates to postulate that: "transsulfuration is the predominate, perhaps exclusive, pathway of homocysteine biosynthesis in the plant kingdom" (Giovanelli et al., 1979).

Cystathionine can serve as a precursor either of methionine or cysteine; the reaction product depends solely on which side of the sulfur atom the molecule is cleaved. Cystathionine β-lyase, which catalyzes the conversion of cystathionine to homocysteine, pyruvic acid, and ammonia, is distributed widely in higher plants (Giovanelli and Mudd, 1971; Datko et al., 1974) and has been extensively purified from spinach leaves (Giovanelli and Mudd, 1971). On the other hand, γ-cleavage yielding cysteine, 2-oxobutyrate, and ammonia is of little consequence to higher plants; it amounts to less than 0.5% of the β-cleavage rate (Giovanelli and Mudd, 1971). In contrast, mammals and fungi have an active γ-cleavage system and can synthesize cysteine from cystathionine (Fig. 35). Higher plants are properly viewed as utilizing β-elimination to convert cystathionine to homocysteine with little evidence existing for significant cysteine production from cystathionine via γ-elimination. Certain bacteria and fungi share with higher plants the ability to exhibit β-cleavage of cystathionine. Rhizobitoxine-treated plants acquire massive amounts of cystathionine, and this accumulation correlates with inhibited β-cystathionase activity (Giovanelli et al., 1971, 1973). Rhizobitoxine is a potent, active site-directed, irreversible inhibitor of the β-cleaving enzyme: cystathionine-β-lyase. This observation further corroborates the importance of the β-elimination reaction in higher plant homocysteine formation (Fig. 35).

Transformation of homocysteine to methionine requires a methyl group donor. In higher plants, probably in all plants, it is S-adenosylmethionine which is formed by methionine adenosyltransferase:

$$\text{ATP} + \text{L-methionine} \rightarrow \text{S-adenosyl-L-methionine} + \text{PP}_i + \text{P}_i$$

This important enzyme has been obtained from several higher plants including barley (Mudd, 1960) and pea (Clandinin and Cossins, 1974). Germinating pea seeds actively synthesize both S-adenosylmethionine and S-adenosylhomocysteine (Dodd and Cossins, 1969). These workers concluded, however, that S-adenosylmethionine was not formed from S-adenosylhomocysteine but was probably derived directly from methionine.

Other methyl group donors are known from higher plants, e.g., S-methylmethionine and methylcysteine but apparently they are of relatively minor importance as compared to S-adenosylmethionine (see Giovanelli *et al.*, 1980). On the other hand, S-methylmethionine is a major metabolite of jack bean, *Canavalia ensiformis* (Greene and Davis, 1960). There is some evidence that methylcysteine functions as a C_1 source after its conversion to formic acid via methylmercaptan. Formic acid is reduced subsequently and incorporated into a receptor molecule (see Doney and Thompson, 1971). In addition, methylmercaptan can function in lieu of sulfide in sulfhydration reactions (Thompson and Moore, 1967; Giovanelli and Mudd, 1968). S-Methylmethionine functions as a methionine reservoir in the senescing flower tissue of *Ipomoea tricolor*. It yields methionine during aging in its function as a methyl group donor to homocysteine (Suttle and Kende, 1980). S-Methylmethionine is also a natural product in the mature petals of *Tradescantia* but the rise in available methionine noted as the petal undergoes senescence results from protein degradation not S-methylmethionine.

Methionine can also react with serine, in a methylthio group transfer reaction, to yield S-methylcysteine which can be converted to its sulfoxide. S-Methylcysteine sulfoxide is a major sulfur reservoir for sulfur-containing amino acid production (Mae *et al.*, 1971). Mazelis (1963) has isolated an enzyme from cruciferous plants that degrades this sulfoxide to pyruvate, ammonia, and methanethiol sulfinate. In addition to its role in protein synthesis, methionine is also the precursor of ethylene, a plant hormone that functions in the ripening process and regulates other aspects of plant growth and development. Ethylene synthesis involves ATP-dependent adenylation of methionine to S-adenosylmethionine;

the latter is cleaved to 1-aminocyclopropyl-1-carboxylic acid and methyl-
thioadenosine prior to ethylene formation (Adams and Yang, 1979; Boller *et al.*, 1979). Pyridoxal phosphate functions in these metabolic con-
versions and canaline has been shown to be a potent inhibitor of
ethylene production from methionine (Murr and Yang, 1975).

Finally, Konze and Kende (1979) have reported that methionine
adenosyltransferase from *Ipomoea tricolor* is active catalytically with
selenomethionine (the V_{max} is twice that noted with methionine itself!)
in the formation of ethylene. Experimental evidence for the intermediary
formation of S-adenosylmethionine in the conversion of methionine to
ethylene was also provided (Konze and Kende, 1979).

2. Lysine

Several nonprotein amino acids function as intermediates in lysine
biosynthesis from aspartic acid. Aspartic acid is initially activated by
ATP and subsequently reduced to 3-aspartylsemialdehyde. These early
reactions are common to homoserine synthesis, but in lysine production
the semialdehyde reacts with pyruvic acid in an aldol condensation with
the concomitant loss of water to yield 2,3-dihydrodipicolinic acid (Fig.
36). This compound is subsequently reduced, the ring opened, and the
amino group protected by succinylation. It is possible that higher plants
use other acylation agents than succinyl-CoA; prokaryotes are known
to employ acetyl-CoA as well. An additional amino group is introduced by transamination to complete formation of N^6-succinyl-2,6-
diaminopimelic acid. This compound is then deacylated by hydrolysis
and the carboxyl group removed in a decarboxylation to form lysine. It is
significant that except for diaminopimelic acid decarboxylase and dihy-
drodipicolinic acid synthase, the putative enzymes of this pathway have
not been isolated from a higher plant source (Mazelis *et al.*, 1977). Thus,
their occurrence is taken by analogy with established microbial enzymes
that catalyze these reactions (see Bryan, 1980).

Both the *meso* and the L,L forms of 2,6-diaminopimelic acid are natural
products. Prokaryotes have a racemase that produces the *meso* isomer
which is converted either to lysine or used directly in cell wall production. It is not yet established if the 2,6-diaminopimelic acid decarboxylase of higher plants discriminates absolutely between the L,L and
meso isomers. This enzyme from *Lemna* (Shimura and Vogel, 1966) and
corn (Sodek, 1978) are active only with the *meso* form of this non-protein
amino acid.

The above pathway (Fig. 36) is distinctive from the aminoadipic acid
pathway of fungi in which 2-oxoadipic acid is transaminated to

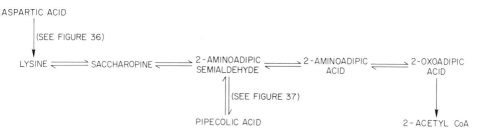

2-aminoadipic acid, converted to its semialdehyde, and then sac-charopine prior to lysine. There are occasional claims that this fungal pathway operates in higher plants but they have never been substan-tiated. Saccharopine and 2-aminoadipic acid, however, may be compo-nents of higher plant lysine degradation (Sørensen, 1976). This possibil-ity is also suggested by the finding of Møller (1976) that barley converts lysine to saccharopine.

Saccharopine forms 2-amino-6-oxohexanoic acid (2-aminoadipic semialdehyde) which potentially at least can yield L-pipecolic acid via Δ^1-piperidine-6-carboxylic acid. Analysis with barley failed to provide convincing evidence of pipecolic acid production from L-[U-^{14}C]lysine. Instead 2-amino-6-oxohexanoic acid (2-aminoadipic semialdehyde)

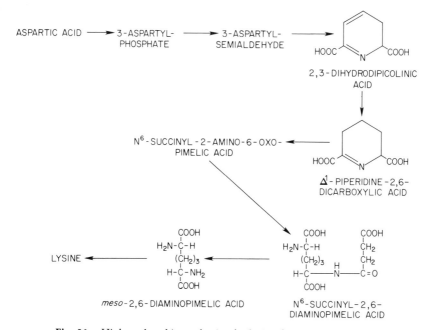

Fig. 36.   Higher plant biosynthesis of L-lysine from L-aspartic acid.

yielded 2-aminoadipic acid, 2-oxoadipic acid, and eventually was found to degrade to acetyl-CoA (Møller, 1976). L-Lysine is not transformed to pipecolic acid but rather degraded to acetyl CoA utilizing a reaction pathway that is common to lysine formation by fungi.

As is often the case with putative metabolic pathways in higher plants, key enzymes and/or metabolites often have not been isolated and characterized. A critical component of lysine production, namely, 2,6-diaminopimelic acid is known only from one plant source-pine pollen. This determination relied solely on evidence gathered by paper partition chromatography. A deliberated search, instituted for this amino acid with the pollen of white pine, *Pinus strobus,* failed to reveal this natural product (Larsen and Norris, 1976). It has still not been obtained unequivocally from plant material but whether this reflects a simple lack of organized searching or its rapid turnover to lysine is not known.

Transformation of lysine to pipecolic acid is an example of the role of lysine in secondary metabolic reactions (Fig. 37). These reactions can proceed with a transamination reaction, cyclization to Δ^1-piperidine-2-carboxylic acid and reduction to pipecolic acid (route I) or by an initial formation of a semialdehyde and its conversion to Δ^1-piperidine-6-carboxylic acid prior to pipecolic acid (route II). Gupta and Spenser (1970) sought to differentiate between these routes by isolating newly synthesized pipecolic acid from plant material provided radioactive lysine having a ^{3}H (β-carbon) to ^{14}C (ω-carbon) ratio of 13.2 ± 0.4. In the isolated product, the label ratio fell to only 0.6 which means that the tritium associated with the β-carbon was largely lost. This finding is consistent with lysine conversion to pipecolic acid via route I. *Pisum sativum* and *Phaseolus radiatus* mediate a NAD(P)H-dependent reduction of Δ^1-piperidine-2-carboxylic acid to L-pipecolic acid (Meister *et al.,* 1957). On the other hand, Fowden's (1960) labeled-precursor-feeding

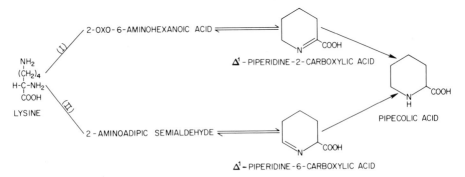

Fig. 37.   Alternative possibilities of higher plant production of L-pipecolic acid.

studies with *Acacia phyllodea* provided evidence for the operation of route II in this legume.

Mammals and fungi can also produce pipecolic acid but D-lysine is the precursor in the former while the latter employs L-lysine. Mammals can use L-lysine but saccharopine and 2-aminoadipic acid rather than pipecolic acid are produced; quantitatively, this is much less active than the mammalian D-lysine pathway (Grove *et al.*, 1969). In this regard, the D-isomeride of lysine is the more effective precursor of L-pipecolic acid in ryegrass and corn (Aldag and Young, 1970) as well as tobacco (Gilbertson, 1972). This observation has been confirmed in tobacco and *Sedum acre* (Leistner *et al.*, 1973) and by the previously mentioned studies of Møller (1976) in that DL-[1-^{14}C]lysine but not L-[U-^{14}C]lysine served as a precursor for the synthesis of radioactive pipecolic acid.

When honey locust, *Gleditsia triacanthos* is provided ^{14}C-containing 5-hydroxy-DL-lysine, 90% of the ^{14}C comes to be partitioned about equally between *cis*- and *trans*-5-hydroxypipecolic acid. The possibility of isomerization as an artifact of the isolation process was discounted by the authors (Thompson and Morris, 1968). This finding of both isomers in a single plant species is significant in two regards. First, it demonstrates that hydroxylated pipecolic acid derivatives may also arise from hydroxylysine. Second, most derivatives of pipecolic acid obtained initially from plant materials were of the *trans* configuration. The observation of Schenk and Schütte (1963) of *cis* and *trans* forms of 4-hydroxypipecolic acid and the confirmatory findings that both isomers of 5-hydroxypipecolic acid are also in *Gymnocladus dioicus* (Kentucky coffee-tree) (Despontin *et al.*, 1977) suggest that these isomeric forms are produced more universally by higher plants than was thought originally.

C. THE GLUTAMIC ACID FAMILY

1. Glutamic Acid Derivatives

Glutamic acid and its amide are the nucleus for a group of derivatives that constitute the largest amalgam of nonprotein amino acids found in higher plants. Many are known to coexist either in the same or closely related plants. This has given rise to metabolic schemes interrelating members of this family although experimental evidence substantiating these reactions is not always complete. While the archetypal compound of this hierarchy is glutamic acid, leucine is actually of considerable importance since it is a prime precursor for many derivatives of glutamic acid (Fig. 38). Glutamic acid formation occurs principally by transamination of 2-oxoglutaric acid provided via Krebs cycle reactions as well as by the action of GOGAT. There is ample experimental proof for a parallel series of reactions forming the oxo-derivative of 2-aminoadipic acid, the higher homolog of glutamic acid (Strassman and Weinhouse, 1953).

In one of the earliest examinations of glutamic acid derivative formation, Linko and Virtanen (1958) provided $[1\text{-}^{14}C]$- and $[2\text{-}^{14}C]$pyruvic acid to the fern *Asplenium septentrionale* and to *Phlox decussata*, a higher plant. Their experimental results suggested that pyruvic acid underwent an aldol condensation; the transaminated reaction product is 4-hydroxy-4-methylglutamic acid. This finding supported the belief that the abundance of 4-substituted derivatives results from an aldol condensation between such molecules as pyruvic acid or pyruvic acid and another oxo-containing compound, e.g., glyoxylic acid prior to transamination to the amino acid.

This question was readdressed in a study using *Gleditsia triacanthos* which showed only slight incorporation of radioactive carbon from $[1\text{-}^{14}C]$pyruvic acid into the α-carboxyl group. This finding precluded direct pyruvic acid condensation in accounting for the formation of these C_6 compounds, for such a reaction would require an equivalent distribution of the label between the two newly formed carboxyl groups. Alternatively, it was proposed that an aldolase-type exchange reaction involving an enzyme–pyruvate complex might be responsible for label incorporation into substituted glutamic acid. Peterson and Fowden (1972) stated:

> The alternative biosynthetic hypothesis invoking the substitution of a C_1 unit on the carbon chain of glutamic acid was excluded by experiments in which $[1\text{-}^{14}C]$glutamic acid and [*methyl*-^{14}C]methionine were supplied separately to growing seedlings. Certainly, glutamic acid was not converted directly into γ-methylglutamic acid; only 10.5% of the total label incorporated into the latter amino acid was located in the α-carboxyl carbon, a figure totally inconsistent with such a conversion. Likewise, the absence of label in any of the γ-substituted glutamic acids after administration of $[^{14}C]$methionine gave no support to the idea of a C_1 group transfer reaction.

When $[1\text{-}^{14}C]$leucine was provided to seedlings, the label in the resulting 4-methylglutamic acid was located almost entirely in the α-COOH group. (2S,4S)-4-Hydroxy-4-methylglutamic acid also became radioactive, probably by conversion of the 4-methylated derivative. Hydroxyleucine may be a reaction intermediate between leucine and 4-methyleneglutamic acid. Further credence to these concepts was provided when $[2\text{-}^{14}C]erythro$-4-methylglutamic acid and $[2\text{-}^{14}C]threo$-4-methylglutamic acid produced the (2S,4S) and (2S,4R) isomers, respectively, of 4 hydroxy-4-methylglutamic acid. The reversal of this reaction, i.e., reduction of 4-hydroxy-4-methylglutamic acid to the 4-methyl derivative, does not appear to occur.

With the above labeled precursors, detectable ^{14}C was not found in 4-methyleneglutamic acid or its amide; thus, conversion of 4-methylglutamic acid to 4-hydroxy-4-methyl-glutamic acid was taken to be a direct reaction. Perhaps, an independent pathway operates in the production of these methylene derivatives. Peterson and Fowden (1972) did offer a plausible explanation for the lack of label transfer to 4-methyleneglutamic acid and its amide that form to an appreciable extent in the germinated seed. Referring to these compounds, they stated that they may: "originate from an unidentified product formed in the cotyledons during seed germination; this substance may be confined to a subcellular metabolic pool that remains distinct from the labelled exogenous precursors supplied in our experiment."

3-Hydroxy-4-methylglutamic acid isomerides have been isolated from a companion leguminous tree species, *Gymnocladus dioicus* (Kentucky coffee tree). A careful probe of their absolute configuration revealed a (2S,3S,4R) and (2S,3R,4R) configuration (Dardenne *et al.*, 1972). Preliminary evidence exists for a synthetic reaction from leucine in which

Leucine → 3-hydroxy-4-methylglutamic acid → 3-hydroxy-4-methyleneglutamic acid

Gleditsia caspica, another leguminous tree species, stores (2S,4R)-4-methylglutamic acid and (2S,3S,4R)-3-hydroxy-4-methylglutamic acid (Dardenne *et al.*, 1974). These results provide added evidence that the pathways emanating from leucine may be common to other plant species (Fig. 38).

An interesting derivative of glutamic acid, an amino acid glycoside in which the carbohydrate moiety is linked to an aliphatic hydroxyl group, is (2S,4R)-4-(β-D-galactopyranosyloxy)-4-isobutylglutamic acid (Larsen *et al.*, 1973). This natural product is found only in the flower, no evidence for its presence in vegetative cells has been obtained. Its concentration appears to be light dependent and it is believed to attain upon full illumination an internal level comparable to that of many protein amino acids. Among the other uncommon derivatives are the 4-ethylidene and 4-propylidene forms. Most of the glutamic acid derivatives are substituted on C-4 but some 3-substituted components are formed by higher plants (Fig. 38).

In rat, hydroxyproline is catabolized to Δ^1-pyrroline-3-hydroxy-5-carboxylic acid and then oxidized to 4-hydroxyglutamic acid (Dekker

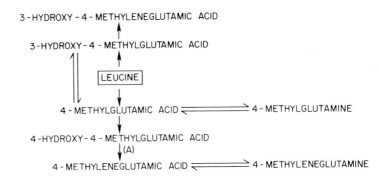

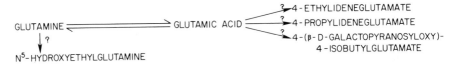

Fig. 38. Some metabolic reactions of the glutamic acid family of nonprotein amino acids. (A) The 4-methylene derivatives may form independently of 4-methylglutamic acid. Question marks denote pathways that have not been elucidated.

and Maitra, 1962). There is no real evidence at present for higher plant production of hydroxylated glutamic acid compounds from hydroxy-proline but this possible alternative route cannot be discounted.

2. Ornithine and Its Derivatives

a. Ornithine

Ornithine is distributed ubiquitously by virtue of its role in arginine biosynthesis which is achieved by the Krebs–Henseleit ornithine–urea cycle reactions or a transamidination in which an amidino group is fused to ornithine. The occurrence of an ornithine–urea cycle in higher plants had been viewed sceptically but this has been dissipated largely by two lines of experimental findings.* The first consists of confirmation that the cycle constituents are higher plant natural products. This is typified by the findings of Kasting and Delwich (1957, 1958) that ornithine, citrul-

*The arguments espoused for considering the reactions of canavanine metabolism as having a synthetic and degradative phase rather than reflecting a functional canaline–urea cycle have been presented (see p. 99). These arguments have equal validity in consider-ing the question of a functional ornithine–urea cycle for higher plants biosynthesis of arginine as compared to a synthetic phase of arginine production from glutamic acid and its degradative reactions via ornithine.

line, and arginine are natural products of wheat, barley, watermelon, and other crops. All four ornithine–urea cycle intermediates are present in 14-day-old *Vicia faba*. When ^{14}C-labeled arginine is vacuum infiltrated into 10-day-old cotyledons, argininosuccinic acid, ornithine, and citrulline pools acquire the administered label (Barber and Boulter, 1963). A relevant review by Naylor (1959) bears on the question of ornithine–urea cycle constituents in higher plants.

The second line of evidence emanates from the established presence of the prerequisite enzymes. Ornithine carbamoyltransferase and arginase are the most active and thoroughly studied of these enzymes. Far fewer investigations have been conducted with argininosuccinic acid lyase or synthetase but all the required enzymes have been found in *Pisum sativum* (Shargool, 1971; Shargool and Cossins, 1968), *Canavalia ensiformis* (Rosenthal, 1972), and most recently in *Vitis vinifera* (Roubelakis and Kliewer, 1978a,b,c).

Ornithine formation in higher plants can also involve a reaction sequence emanating from glutamic acid (Fig. 39). In the initial step, acetyl-CoA serves in the production of N-acetylglutamic acid; the N-acetyltransferase fostering this reaction has been isolated from several higher plants (Morris and Thompson, 1977). The γ-COOH group is then activated in an ATP-dependent reaction in which 4-glutamyl phosphate is formed and converted to the corresponding semialdehyde. Cyclization of the semialdehyde is prevented by prior α-NH₂ group protection via acetylation. In proline biosynthesis, this group is not protected and this permits spontaneous cyclization of glutamic acid semialdehyde to Δ^1-pyrroline-5-carboxylic acid. N-Acetyl-4-glutamylsemialdehyde is

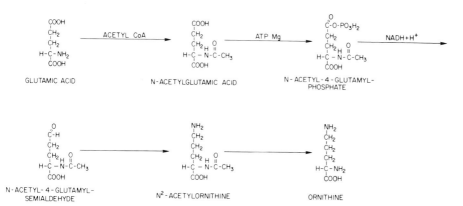

Fig. 39. Nonprotein amino acids functioning in ornithine production from glutamic acid.

converted to N^2-acetylornithine in a transamination reaction in which glutamic acid can function as the amino group donor; reduced NAD(P) or NAD can provide the required reducing power. Ornithine results from deacetylation of N^2-acetylornithine and the acetyl group is conserved since it can be reacted directly with glutamic acid by a transacetylase (Morris and Thompson, 1977):

$$N^2\text{-Acetylornithine} + \text{glutamate} \rightarrow N\text{-acetylglutamate} + \text{ornithine}$$

Transacetylation represents an alternative to direct acetylation of glutamic acid by acetyl-CoA. In *Beta vulgaris* (beet), the K_m for acetylornithine is 0.2 mM as compared to 2.5 mM for acetyl-CoA. Additionally, the acetylornithine concentration in beet is greater than that of acetyl-CoA. These factors form the basis for the assertion that in certain higher plants the transacetylation pathway may predominate over that of acetylation in the formation of N-acetylglutamate (Thompson, 1980).

Glutamic acid, after its conversion to 4-glutamylsemialdehyde, can be transaminated directly to ornithine; this results in more direct ornithine production (see Fig. 43, p. 188). As the ornithine to 4-glutamylsemialdehyde reaction is reversible, it has come to be accepted widely that ornithine can also yield proline via this semialdehyde by way of Δ^1-pyrroline-5-carboxylic acid.

It is of interest that recent tracer experiments with several green plants have suggested that transformation of ornithine to proline via 4-glutamylsemialdehyde is a minor pathway (Mestichelli *et al.*, 1979). These workers discovered that ornithine is metabolized to proline with the loss of the α-NH$_2$ rather than the δ-NH$_2$ group. This involves the sequential formation of 2-oxo-5-aminopentanoic acid and Δ^1-pyrroline-2-carboxylic acid (see Fig. 43). Dual labeling studies with [2-^{3}H,5-^{14}C]ornithine and [5-^{3}H,5-^{14}C]ornithine revealed that the ratio of ^{3}H to ^{14}C in proline decreased only with [2-^{3}H,5-^{14}C]ornithine. On the other hand, there is a mass of evidence from many higher plants including *Lupinus angustifolius* and *Vigna radiata* as well as *Lathyrus latifolius* that ornithine and 2-oxoglutaric acid react to form glutamic acid and 4-glutamylsemialdehyde (Hasse *et al.*, 1967; Seneviratne and Fowden, 1968).

The production of ornithine from glutamic acid interrelates the ornithine–urea cycle with the reactions of the glutamic acid family. Morris and Thompson (1977) have determined that arginine can inhibit the initial acetylation of glutamic acid as well as its subsequent phosphorylation. No evidence, however, was obtained for end-product repression of the committed step. These findings confirm an earlier study with *Chlorella vulgaris* (Morris and Thompson, 1975) and focus on the control points in arginine regulation of glutamic acid support of ornithine synthesis.

b. *Acetylated Derivatives*

Higher plants not only synthesize the N^2-acetylated derivative of ornithine, a reaction intermediate in ornithine production from glutamic acid, but also N^5-acetyl-L-ornithine (Brown and Fowden, 1966). N^5-Acetylornithine was isolated initially by Menske (1937) from the root tissues of *Corydalis ochotensis* and *C. bulbosa* where it can constitute 10% of the fresh weight. Menske's assertion that this newly discovered natural product represented the N^5- rather than the N^2-acetylated derivative was based solely on its ninhydrin responsiveness and sweet taste! It was an entirely fortuitous surmise since both the N^2- and N^5-acetyl derivatives are ninhydrin-positive.

N^5-Acetylornithine was obtained subsequently from the fern *Asplenium nidus* (Virtanen and Linko, 1955), and shown to be widely dispersed among the Fumariaceae (Fowden, 1958). The identity of the compound isolated by Menske was not established properly until the efforts of Brown and Fowden (1966) utilizing materials obtained from *Onobrychis viciifolia* (sainfoil).

Feeding studies utilizing N^5-acetyl-DL-[2-^{14}C]ornithine and DL-[2-^{14}C]ornithine with *Phaseolus aureus* and *Onobrychis viciifolia* have been conducted. *Phaseolus aureus* converts free ornithine to arginine and transfers the appropriate carbon atoms from ornithine–urea cycle constituents to proline, glutamic acid, and amino acid amides. By contrast, in *O. viciifolia* nearly 80% of the recovered ornithine label exists as N^5-acetylornithine (Brown and Fowden, 1966). There is little movement of the acetylated derivative back to ornithine.

Study of the acetylation process in *Onobrychis* indicates that N-acetylglutamic acid is 4 times more effective an acetylating agent than is acetyl-CoA. Neither coenzyme A nor acetyl-CoA stimulates the acetyl transfer ability of potential donor molecules. This suggests a direct transfer mechanism which does not involve the intervention or intermediary formation of acetyl-CoA. N^2-Acetylornithine functions as an acetylating agent for the nitrogen atom linked to C-5 but amino acids bearing their acetyl group in a position other than the α-amino group cannot contribute effectively to the biosynthesis of N^5-acetylornithine (Brown and Fowden, 1966).

N^5-Acetylornithine also supports reversible production of a novel metabolite, namely, 5-acetoamino-2-hydroxypentanoic acid in *Nicotinia tabacum* (Noma *et al.*, 1978). It was suggested that this novel compound may be oxidized to N^5-acetyl-2-oxo-5-aminopentanoic acid which when deacylated produces 2-oxo-5-aminopentanoic acid. The latter compound cyclizes to Δ^1-pyrroline-2-carboxylic acid (see Fig. 43), which can yield proline. If so, this pathway would represent a novel biosynthetic route from ornithine to proline.

3. 5-Aminolevulinic Acid

The studies of Granick, conducted during the 1940s with mutant chlorophyll-deficient strains of *Chlorella*, provided evidence that heme and chlorophyll share certain common biosynthetic reactions. Other natural products radiating from these reactions include vitamin B_{12}, various bilins, the prosthetic group of certain nitrite and sulfite reductases, accessory photosynthetic pigments of prokaryotic and eukaryotic algae, and phytochrome's chromophore (Beale, 1978). 5-Aminolevulinic acid serves as precursor in all these reactions; amalgamation of 8 molecules of this non-protein amino acid forms the tetrapyrrole ringed structures of protoporphyrin IX, a progenitor of heme and chlorophyll.

In certain protistans, fungi, and mammals, 5-aminolevulinic acid is produced by condensation of glycine and succinyl-CoA in a pyridoxal phosphate-dependent reaction. This reaction, driven by 5-aminolevulinic acid synthetase, involves the decarboxylation of an enzyme-bound reaction intermediate, 2-amino-3-oxoadipic acid (Laghai and Jordan, 1977).

$$
\begin{array}{c}
\text{COOH} \\
|\ \\
\text{CH}_2 \\
|\ \\
\text{CH}_2 \\
|\ \\
\text{C}=\text{O} \\
|\ \\
\text{SCoA}
\end{array}
\quad\longrightarrow\quad
\begin{array}{c}
\text{COOH} \\
|\ \\
\text{CH}_2 \\
|\ \\
\text{NH}_2
\end{array}
\quad\longrightarrow\quad
\begin{array}{c}
\text{COOH} \\
|\ \\
\text{CH}_2 \\
|\ \\
\text{CH}_2 \\
|\ \\
\text{C}=\text{O} \\
|\ \\
\text{H}-\text{C}-\text{NH}_2 \\
|\ \\
\text{COOH}
\end{array}
\quad\xrightarrow{\ \text{CO}_2\ }\quad
\begin{array}{c}
\text{COOH} \\
|\ \\
\text{CH}_2 \\
|\ \\
\text{CH}_2 \\
|\ \\
\text{C}=\text{O} \\
|\ \\
\text{CH}_2 \\
|\ \\
\text{NH}_2
\end{array}
$$

| Succinyl | Glycine | 2-Amino-3- | 5-Amino- |
| CoA | | oxoadipic acid | levulinic acid |

The requisite enzymes also occur in such distinctive sources as avian erythrocytes and the endoplasmic reticulum of liver cells but are not demonstrated readily in higher plants. Alternative biosynthetic pathways may exist for plant production of this nonprotein amino acid (Fig. 40). Whatever the actual synthetic route, it is not only a major carbon and nitrogen contributor to tetrapyrrole production but also a significant regulatory element in chlorophyll biosynthesis (Beale and Castelfranco, 1974). As such, 5-aminolevulinic acid is a prominent metabolic intermediate of all chlorophyll-employing organisms (see Beale, 1978, for additional information).

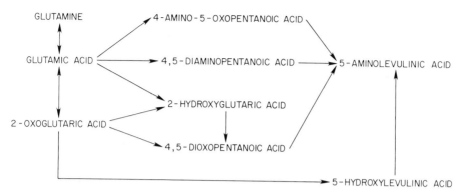

Fig. 40. 5-Aminolevulinic acid formation from glutamine and glutamic acid. After Beale (1978).

D. AROMATIC AMINO ACIDS

Some consideration of aromatic amino acid biosynthesis is desirable before integrating the aromatic nonprotein amino acids into this metabolic matrix. Production of these particular constituents commences with the condensation of phosphoenolpyruvate and erythrose-4-phosphate to eventually build 3-phosphoshikimic acid and then chorismic acid. The latter compound is a major intermediary of aromatic amino acid formation since it occupies a key branch point in phenylalanine plus tyrosine or tryptophan production (Fig. 41). Rearrangement of the enolpyruvyl side chain of chorismic acid results in conversion to prephenic acid (I, Fig. 41) which is decarboxylated prior to transamination to phenylalanine (II, Fig. 41). Alternatively, prephenic acid can function in tyrosine production via 4-hydroxyphenylpyruvic acid (III, Fig. 41). In this case, retention of the OH group of prephenic acid is of critical importance in tyrosine production since evidence of significant plant hydroxylation of phenylalanine to tyrosine is lacking. For example, administration of [^{14}C]phenylalanine to *Vicia faba* fails to produce labeled tyrosine. *De novo* tyrosine synthesis does not occur from polyacetate or by any other known reaction sequence involving acetate (Griffith and Conn, 1973) (see also p.90).

Jensen and collaborators (Jensen and Pierson, 1975; Patel *et al.*, 1977) presented evidence creating the possibility in prokaryotes that prephenic acid is transaminated directly to produce pretyrosine (IV, Fig. 41). This pretyrosine pathway (IV) is taken to represent an early evolutionary mechanism and is associated with such protistans as blue-green

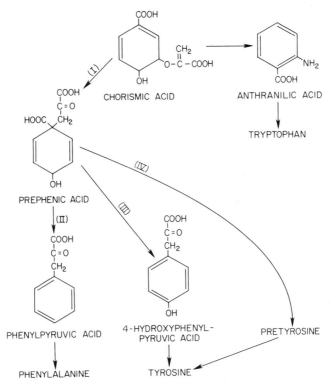

Fig. 41.   Protein amino acid synthesis from chorismic acid.

algae. Pathway III in contrast is believed to represent a highly evolved pathway and one operative presently in such enterics as *E. coli* (Jensen and Pierson, 1975).

Examination of tyrosine production in the mung bean, *Vigna radiata* (Rubin and Jensen, 1979), led to the isolation of prephenate dehydrogenase which converts prephenic acid to 4-hydroxyphenylpyruvic acid. A 4-hydroxyphenylpyruvate aminotransferase, also obtained from this legume, is responsible for the conversion of 4-hydroxyphenylpyruvate to tyrosine (III, Fig. 41). Interestingly, prephenic acid can also be converted to pretyrosine by prephenate aminotransferase and then to tyrosine by virtue of the catalytic action of pretyrosine dehydrogenase (IV, Fig. 41). Evidence was presented for a single 52,000 dalton protein which is responsible for both of the dehydrogenase activities culminating in tyrosine biosynthesis. This protein is a highly unusual example of a single enzyme that functions in the biosynthesis of a particular metabolite by two divergent pathways (Rubin and Jensen, 1979).

Another distinctive higher plant modification of these basic metabolic interactions has been suggested from the work of Larsen and associates (1975) who proposed that chorismic acid gives rise to isochorismic acid prior to isoprephenic acid (Fig. 42). Once formed, many nonprotein amino acids can emanate from these precursor compounds. Isoprephenic acid can yield 3-(3-carboxyphenyl)alanine (*A*) or 3-(3-carboxy-4-hydroxyphenyl)alanine (*C*) by transamination from their corresponding phenylpyruvic acid derivatives; the latter derivatives are not depicted in Fig. 42. These derivatives of phenylalanine are rendered radioactive in feeding studies involving stereospecifically labeled shikimic acid. These studies served as the basis for the contention that chorismic acid conversion to isochorismic acid and isoprephenic acid are requisite reactions to aromatic nonprotein amino acid formation (Fig. 42). Hydroxylation of (*A*) to (*C*) has been demonstrated in *Reseda lutea* and *R. odorata* (Larsen and Sørensen, 1968) but it is less important quantitatively than their direct production from isoprephenic acid.

The biosynthetic reactions involving certain carboxylated aromatic derivatives shown in Fig. 42 have been analyzed in *R. lutea* and *R. odorata*. Larsen (1967) administered [*U*-¹⁴C]shikimic acid and isolated radioactive 3-(3-carboxyphenyl)alanine (*A*), 3-(3-carboxy-4-hydroxyphenyl)glycine (*C*), and 3-(3-carboxy-4-hydroxyphenyl)alanine (*D*). Degradation of these molecules revealed that their aromatic ring was obtained from the ring of shikimic acid. The aromatic carboxyl group of the newly formed compounds was also provided by shikimic acid. Tyrosine was shown to yield 3-(3-carboxy-4-hydroxyphenyl)alanine but significantly phenylalanine did not produce 3-(3-carboxyphenyl)alanine.

Not only are the (*A*) and (*C*) derivatives of phenylalanine produced from ¹⁴C-shikimic acid, so too are the corresponding 3-carboxy (*B*) and

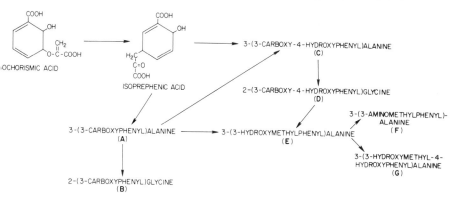

Fig. 42. Aromatic nonprotein amino acid biosynthesis in higher plants.

3-carboxy-4-hydroxy (D) derivatives of phenylglycine. These four aromatic compounds, i.e., (A), (B), (C), and (D), coexist in certain species of Cruciferae, Resedaceae, and Iridaceae (Kjaer and Larsen, 1974).

The above glycine-containing compounds may form by chain shortening of their alanine-containing counterparts. Morris and Thompson (1964, 1965) have demonstrated appreciable conversion of ^{14}C-labeled 3-(3-carboxyphenyl)alanine (A) to radioactive 2-(3-carboxyphenyl)glycine (B) in Iris leaves. A similar approach was employed in analyzing the conversion of 3-(3-carboxy-4-hydroxyphenyl)alanine (C) to its corresponding glycine-containing derivative (D). Product recovery of the applied radioactivity was only 0.5% with Iris but it increased to an average incorporation rate of 1.26% with Reseda odorata. These chain-shortening reactions involve the intermediary formation of the corresponding mandelic and glyoxylic acid constituents (Larsen and Wieczorkowska, 1975). Details of the decarboxylation step are not available but these reactions are not unique to Resdea since 2-(3-hydroxyphenyl)glycine forms from 3-(3-hydroxyphenyl)alanine in Euphorbia helioscopia (Müller and Schütte, 1971).

It is reasonable to propose that 3-(3-hydroxymethylphenyl)alanine (E), produced by reduction of 3-(3-carboxyphenyl)alanine (A), may in turn be a precursor for 3-(3-hydroxymethyl-4-hydroxyphenyl)alanine (G). These reactions are believed to operate in Caesalpinia from which these nonprotein amino acids, i.e., (A), (E), and (G), have been isolated. More recently, 3-(3-hydroxymethylphenyl)alanine (E) and 3-(3-carboxyphenyl)alanine (A) were obtained from Iris sanguinea and I. sibirica (Larsen et al., 1978).

It is of interest that Vigna vexillata produces 4-aminophenylalanine directly from shikimic acid without preformation of phenylalanine or tyrosine (Dardenne et al., 1975). This observation is in accord with an earlier contention that the compounds depicted in Fig. 42 are produced without direct phenylalanine or tyrosine intervention.

E. HETEROCYCLIC COMPOUNDS

1. O-Acetylserine Function

About 20% of plant nonprotein amino acids are of a heterocyclic nature and these are characteristically very diversified in nature. Many are β-substituted alanines. Extrapolating from an earlier appreciation of serine's role in providing the side chain of tryptophan, it was proposed that serine might also be the source of the β-alanine side chain for other nonprotein amino acids. Evidence exists for this role for serine in the biosynthesis of β-pyrazol-1-ylalanine [164], mimosine [81], and orcylalanine [120]. Giovanelli and Mudd (1967, 1968) subsequently showed that O-acetylserine rather than serine per se provides carbon for the production of certain sulfur amino acids (see p. 162). Indeed, O-acetylserine was 60 times more active than serine in the formation of cysteine by spinach (Giovanelli and Mudd, 1967).

$$H_3C-\overset{\overset{\displaystyle O}{\|}}{C}-O-CH_2-CH(NH_2)COOH$$

O-Acetylserine

Drawing on this information, Murakoshi *et al.* (1972a) combined an extract of *Leucaena,* a legume known to synthesize mimosine readily, with 3,4-dihydroxypyridine and ^{14}C-labeled O-acetylserine; radioactive mimosine was obtained subsequently. Similarly, *Citrullus vulgaris* (watermelon) mediates β-pyrazol-1-ylalanine production from pyrazole and O-acetylserine. Neither serine nor O-phosphoserine functions in this reaction.

Pyrazole　　　　　　　　　　β-Pyrazol-1-ylalanine

Free pyrazole occurs in watermelon but not in several other members of the Cucurbitaceae (LaRue and Child, 1975). The previously described experimental findings have now firmly established the principle that β-substituted alanines can be derived by condensation reactions in which O-acetylserine is the active carbon donor. There is little evidence that O-phosphoserine or some other serine derivative replaces O-acetylserine as a substrate for these heterocyclic amino acid biosyntheses (Kjaer and Larsen, 1974). Other possible O-acetylserine recipients include quisqualic acid [165] and ascorbalamic acid [132]. The former is created by reaction with 3,5-dioxo-1,2,4-oxazolidine and an extract of *Quisqualis indica* (Murakoshi *et al.,* 1974). As such it is the only naturally occurring amino acid containing the 1,2,4-oxadiazolidine ring. Ascorbalamic acid [132] may be distributed widely and there is also the possibility that it is formed by condensation with ascorbic acid. Acidic treatment of ascorbalamic acid yields 3-(2-furoyl)alanine [142]. These two compounds, i.e., [132] and [142], may be related metabolically since they have been shown to coexist both in monocots and dicots.

2. Uracil and Isoxazolinone Derivatives

Growing pea seedlings accumulate increasing levels of two uracil-containing compounds during the initial weeks of growth: willardiine [172] (1-alanyluracil) and its 3-alanyluracil isomeride [147] (Lambein and van Parijs, 1968). A compound putatively designated 5-alanyluracil has been isolated from pea (Brown and Silver, 1966) but NMR and mass spectroscopy by Brown and Mangat (1969) culminated in the revision of its structure to that of 3-alanyluracil. Thus, only two isomers of willardiine are known to be proliferated by higher plants. Reports of the natural occurrence of isowillardiine are limited presently to *Pisum* but willardiine has been obtained from *Acacia willardiana* (Gmelin, 1959), various *Acacia* (Evans *et al.*, 1977), and in other members of the Mimosoideae (Gmelin, 1961; Krauss and Reinbothe, 1973).

Willardiine and its isomer appear to be synthesized from a preformed pyrimidine ring which itself is obtained from the orotic acid pathway. Once again, *O*-acetylserine is more active than free serine in formation of the alanyl side chain. Appreciable [2-^{14}C]uracil and [6-^{14}C]orotic acid are incorporated into willardiine and its isomer by seeds of *Pisum sativum* (Ashworth *et al.*, 1972). The biosynthesis of willardiine and isowillardiine has been considered more recently by Murakoshi *et al.* (1978). These workers reported a K_m for uracil of about 11 mM in the synthesis of willardiine and about 3.3 mM for isowillardiine production. Uracil exhibited marked substrate inhibition in its formation of both of these non-protein amino acids. *O*-Acetyl-L-serine was implicated firmly as the physiological donor of the alanyl side chain (Murakoshi *et al.*, 1978).

Lambein and van Parijs' researches have resulted in the isolation and characterization of many isoxazolinone natural products of which three are non-peptidyl amino acids. These heterocyclic compounds are excluded from the newly ripened seed but make a dramatic appearance with seed growth. The initial members of this group, namely, 3-(isoxazolin-5-one-2-yl)alanine [148] and 3-(2-β-D-glucopyranosyl-isoxazolin-5-one-4-yl)alanine [149], were obtained from pea, *Pisum sativum* (Lambein and van Parijs, 1970). The former compound constitutes about 1 to 2% of the seedling weight. The same two substituted isoxazolin-5-one compounds also occur in *Lathyrus odoratus* along with 2-amino-4-(isoxazolin-5-one-2-yl)butyric acid [150]; the latter is restricted to *Lathyrus* (Lambein and van Parijs, 1974).

The isoxazole-5-one compounds form from free isoxazolin-5-one. In the case of the β-D-glucopyranose derivative [149], this necessitates a condensation with uridine diphosphate glucose. *O*-Acetylserine provides the alanyl side chain. Glucosylation of the ring nitrogen is a prerequisite to the alanylation reaction (Lambein *et al.*, 1976). The simpler compound, namely, 3-(isoxazolin-5-one-2-yl)alanine [148], is formed in a direct condensation of isoxazolin-5-one and *O*-acetylserine. The enzymatic synthesis of [148] by this means has been established in *Pisum* (Murakoshi *et al.*, 1972b). The lathyritic compound [150] has a butyryl side chain but the C_4 unit can be supplied by *O*-acetylhomoserine or *O*-oxalylhomoserine rather than *O*-acetylserine since these acylhomoserine derivatives are natural products of *Lathyrus* (see p. 162). 3-Isoxazolin-5-one-2-yl)alanine has a weak but discernible neurological action in vertebrate neurons.

3. 4-Pyrone Derivatives

In 1959, Hattori and Komamine discovered a novel 4-pyrone derivative from *Stizolobium hassjoo* which was trivially named stizolobic acid [166]. The generic source name is synonymic with *Mucuna*, best known for its production of dopa. An isomeric form with the alanyl side chain at C-3, instead of C-4 as it is in stizolobic acid, has also been obtained and designated stizolobinic acid [167]. Detailed chemical studies indicate that the structure assigned originally to these novel compounds is correct (Senoh *et al.*, 1964). Feeding experiments with potential precursor compounds were published in 1975 (Saito *et al.*, 1975). In this work, DL-[3-^{14}C]dopa and L-[U-^{14}C]tyrosine, and L-[U-^{14}C]phenylalanine all labeled these pyrone derivatives but the incorporation rate varied considerably and was given as 0.6, 0.3, and 0.06%, respectively. An oxidative cleavage of the aromatic ring was postulated but substantive experimental evidence bearing on the question of 4-pyrone synthesis was not provided at that time.

In 1975, Ellis reexamined the matter of stizolobic acid synthesis in young leaves of *Mucuna deeringiana* and concluded that it was formed from dopa by 4,5-extradiol ring cleavage and subsequent cyclization. 2,3-Extradiol ring cleavage was offered as a basis for stizolobinic acid formation. A stizolobic acid synthase and stizolobinic acid synthase were subsequently isolated from etiolated epicotyls of *Stizolobium hassjoo* (Saito and Komamine, 1978). These enzymes appear to have an absolute specificity for 3,4-dihydroxyphenylalanine when tested against various *o*-dihydric phenols, are activated by Zn^{2+}, and employ $NADP^+$ as an oxidizing agent. The reaction sequence culminating in their production involves extradiol ring cleavage, recyclization, and finally a dehydrogenation. A single catalytic protein appears to mediate the entire biosynthetic sequence (Saito and Komamine, 1978).

4. Isoquinolines

Four naturally occurring isoquinolines are presently known. 1-Methyl-3-carboxy-6-hydroxy-1,2,3,4-tetrahydroquinoline [71] was obtained from *Euphorbia myrsinites* while *Mucuna deeringiana* provides a similar compound except for an additional OH group at C-7 [72]. The compound from *Euphorbia myrsinites* was postulated to form in a reaction with *m*-tyrosine and an active C_1 compound such as acetaldehyde (Müller and Schütte, 1968). More recently, these workers showed that ^{14}C-labeled *m*-tyrosine is converted to [71]. This synthesis probably represents a non-enzymatic process involving a Mannich condensation with formaldehyde (Müller and Schütte, 1971). Bell *et al.* (1971) have alluded to the possible synthesis of the latter compound, i.e., [72], from L-dopa but nothing is known of the actual assimilation of the extra carbon atom into the ringed structure. *Mucuna mutisiana* stores a compound similar to [72] except for the lack of a methyl group at C-1 [70]. Finally, 1-carboxy-6,7-dimethoxy-8-hydroxy-1,2,3,4-tetrahydroisoquinoline is known from the peyote cactus [69]. A relatively complex compound has been isolated from *Aleurites fordii* seeds which has three fused ring systems and is named 3-carboxy-1,2,3,4-tetrahydro-3-carboline [137]. It may be formed from tryptophan and formaldehyde and related to the isoquinolines (Okuda *et al.*, 1975).

5. Imino Acids

Of the many imino acids of higher plants, compounds based on the following types of structurally related ringed forms have been studied most thoroughly:

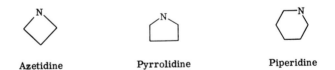

    Azetidine   Pyrrolidine   Piperidine

Azetidine-2-carboxylic acid, representing the simplest member of this group, has been considered in detail elsewhere (section E,2). *Fagus sylvatica* seeds have provided two other members of this family, i.e., N-(3-amino-3-carboxypropyl)azetidine-2-carboxylic acid [134] and N-[N-(3-amino-3-carboxypropyl)-3-amino-3-carboxypropyl]azetidine 2-carboxylic acid [135]. The latter compound is identical to nicotianamine and it has also been obtained from tobacco, *Nicotiana tabacum*. Although these fagaceous compounds can form chemically from a heated aqueous solution of azetidine-2-carboxylic acid, their production is taken to be enzyme-mediated since the latter amino acid is not found in the fruit of this forest species (Kristensen and Larsen, 1974).

A novel function has been suggested for nicotianamine which may also be applicable to other nonprotein amino acids, i.e., the transport of metallic ions (Buděšínský *et al.*, 1980). These workers proposed that nicotianamine possesses a molecular structure amenable to complex formation with iron and functioning in iron transport and/or metabolism and they stated:

> Not only are six functional groups present, necessary for octahedral coordination, but the distances between the groups are also optimal for the formation of chelate rings: Three 5-membered rings formed by the α-amino acid residues and two 6-membered rings formed by the 1,3-diaminopropane moieties. The special location of the oxygen atoms on one side of the complex and the methylene groups and/or the azetidine ring system on the other might play a decisive role in the biological function of the complex.

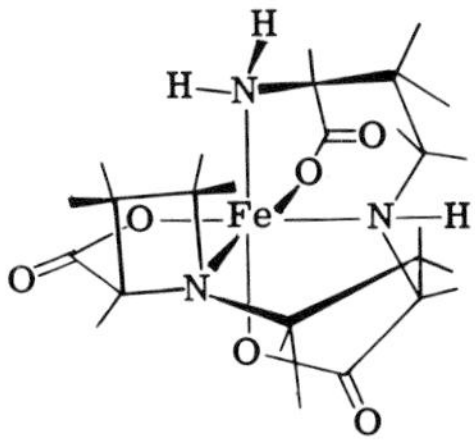

Dreiding model of the iron-nicotianamine complex (taken from Buděšínský *et al.*, 1980).

This suggestion is based on an earlier finding of Takemoto *et al.* (1978) that mugineic acid, N-[3-(3-hydroxy-3-carboxypropylamino)-2-hydroxy-3 carboxypropyl]-azetidine-2-carboxylic acid [81] has iron-chelating ability and this compound may aid in plant uptake of chelated iron via root excretion of this nonprotein amino acid and its reaction with iron prior to plant uptake of the complex. Addition of mugineic acid to the nutrient solution used to sustain rice at pH 7 resulted in increased iron in the leaves. It is interesting to compare the structure of [81] with that of its non-hydroxylated derivative, i.e., [135].

Among the presently known pyrrolidine-based compounds are the *trans*-4-methyl [160] and 4-methylene [163] derivatives of proline. In hygric acid [161], the nitrogen atom employed for ring closure is methylated; 4-hydroxy-N-methylproline [162] is related most closely to it. Proline carboxylated at carbon 4 and 5 [157 and 158a, respectively] as well as 3-hydroxy [93], 4-hydroxy [94], and 4-hydroxymethyl [95] derivatives have also been isolated. The remaining presently known derivatives of proline include cucurbitine [138] and the more complex forms: domoic acid [140] and α-kainic acid [151]. The latter is known to have an isomeric form.

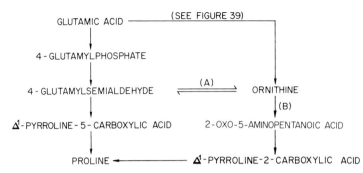

Fig. 43. Metabolic reactions of proline formation from glutamic acid and ornithine. Route (A) represents the established pathway of proline synthesis from ornithine. Recent experimental findings indicate that route (B) may be more important in higher plant metabolism.

Proline biosynthesis proceeds from glutamic acid in a reaction sequence initiated by activation of the dicarboxylic acid with ATP. It is then reduced to the corresponding semialdehyde which cyclizes spontaneously to Δ^1-pyrroline-5-carboxylic acid. The compound is reduced further with NADH + H$^+$ to form proline (Fig. 43). Ornithine can be converted directly to 4-glutamylsemialdehyde, thereby providing an indirect route for proline production.

4-Hydroxyproline is a major constituent of animal collagen and the cell wall of plants. As such it is one of the better known and biologically important nonprotein amino acids. It is generated by posttranslational hydroxylation of peptidyl proline; both plants and animals utilize diatomic oxygen in this reaction. Experiments with $H_2^{18}O$ injected into chick embryo indicate that such isotopic oxygen is not transferred to hydroxyproline. This hydroxylated compound is labeled, however, if the embryo is maintained in an atmosphere of $^{18}O_2$ (Prockop *et al.*, 1962). A reducing agent, Fe^{2+}, and 2-oxoglutarate enhance the reaction. The final factor is a specific requirement in animal systems but plants can use pyruvate or oxaloacetate as well (Sadava and Chrispeels, 1971).

Collagen is an unusual biological molecule formed from three polypeptide chains. These chains have an individual left-handed helical orientation but they intertwine as a right-handed superhelix. Hydroxyproline and proline together constitute about one-third of the residues of this macromolecule. The OH group is particularly important since it stabilizes the triple helix, undoubtedly by hydrogen bonding between chains. In cells in which posttranslational hydroxylation is reduced, the hydroxyproline-deficient collagen unravels at a lower temperature than their normally formed counterparts. This thermal instability results from a diminution in hydrogen bond formation and the subsequent synthesis of amino acid chains having a thermodynamically less stable helical structure. It is now known that the prime function of hydroxyproline is the stabilization of these collagen fibrils (Eyre, 1980).

In one of the earliest studies of plant production of hydroxyproline, Pollard and Steward (1959) established that while carrot cells take up radioactive hydroxyproline, it is not found in synthesized proteins. In contrast, labeled proline is not only incorporated readily into protein but also produces radioactive hydroxyproline. Chase experiments clearly support posttranslational proline alteration in the formation of hydroxyproline by the plant (for further information, see the review by Chrispeels and Sadava, 1974).

Primary plant wall cells contain a glycoprotein rich in hydroxyproline, accounting for 25% of the total residues. The proline-containing polypeptide chains are hydroxylated in the cytoplasm, most of these residues are then glycosylated with arabinose and the newly formed glycoprotein shipped to the cell wall (Chrispeels, 1970). The hydroxylation process occurs after chain release from the polyribosomal complex (Sadava and Chrispeels, 1971). Carrot disks incubated with [^{14}C]proline exhibit a 3 min delay in amino acid uptake. Soon thereafter appreciable radioactive hydroxyproline is noted in the synthesized protein but there is little evidence of residue glycosylation. This process lags about 4 min behind hydroxylation which itself occurs about 4 min after proline incorporation into carrot protein. After 20 min, 85% of the hydroxylated residues are glycosylated (Chrispeels, 1970).

Trans-4-hydroxyproline appears to be a well-dispersed component of plant proteins and it accounts for as much as 20% of the amino acid residues of the primary cell wall. Glycosylation of the hydroxyproline in cell wall proteins is partially active during cell wall alteration as a function of cell enlargement. It is most interesting that carrot enzymes involved in the formation of 4-hydroxyproline are also active with the protocollagen of the chick embryo (Sadava and Chrispeels, 1971).

In the sandal plant, Santalum album, two isomers of hydroxyproline are prominent leaf constituents. Trans-4-hydroxyproline is bound and assumed to form by hydroxylation of peptidyl proline. The cis isomer, however, occurs in the free state and is not formed by epimerization of the bound isomeride (Kuttan and Radhakrishnan, 1970). These authors contend that the cis form is derived from free proline rather than from ornithine or by a condensation reaction such as that between pyruvate and glyoxylate. This observation bears on the question of whether the many hydroxylated derivatives of proline are formed by direct hydroxylation of this protein amino acid. Alternatively, one can imagine reaction sequences involving ornithine that would parallel lysine's precursor role in generating hydroxylated piperidine-containing compounds.

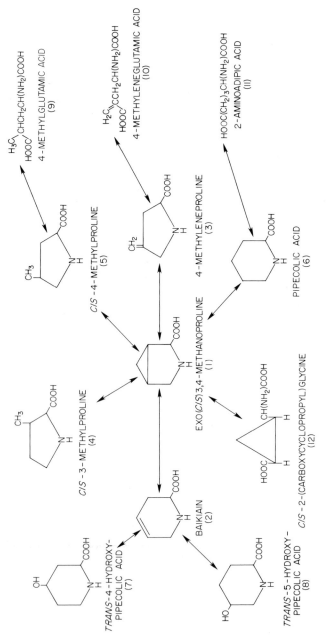

Fig. 44. Putative routes of imino acid production in certain higher plants. Adapted from Fowden (1970).

The piperidine-ringed structures are also quite numerous including a 3-hydroxy [88], both *cis* and *trans* isomers of the 4-hydroxy [89] and 5-hydroxy [90] forms, and the 5-hydroxy-6-methyl [91] and 4,5-di-hydroxy [92] derivatives. A non-hydroxylated form, 4-aminopipecolic acid [131], and the methylated natural product, baikiain [136], also oc-cur. Pipecolic acid formation proceeds from L-lysine in fungi but higher plants preferentially utilize the D-racemate as the starting material. These reactions of pipecolic acid and its hydroxylated forms have been considered elsewhere (see p. 169).

Fowden (1970) has integrated the possible metabolic relationships of this group of compounds into a coherent scheme (Fig. 44). In this model, rearrangement of exo(*cis*)3,4-methanoproline (1) could produce baikiain (2) or 4-methyleneproline (3). Alternatively, reductive cleavages are feasi-ble and these would interrelate *cis*-3- (4) and *cis*-4- (5) methylated pro-line as well as pipecolic acid (6) into these metabolic reactions. Addition to baikiain which is unsaturated may lead to *trans*-4- (7) and/or *trans*-5-hydroxypipecolic acid (8). These relationships also provide for ring structure development from aliphatic saturated and unsaturated com-pounds. For example, 4-methylglutamic acid (9) can feed into *cis*-4-methylproline, and in a parallel manner 4-methyleneglutamic acid (10) into 4-methyleneproline (3), and 2-aminoadipic acid (11) to pipecolic acid (6). Finally, the conversion of *cis*-2-(carboxycyclopropyl)glycine (12) to exocyclic proline (1) may be a key reaction in the formation of certain C_6 imino acids. Catalytic hydrogenation of this exocyclic substituent is known to produce *cis*-3-methylproline.

F. CYCLOPROPYL OR UNSATURATED COMPOUNDS

The researches of Fowden and collaborators are responsible for ex-panding significantly the number of known cyclopropyl and unsatu-rated nonprotein amino acids. Their isolation of 2-(2-methylenecyclo-propyl)glycine from *Litchi chinensis* established this lower homolog of hypoglycin [3-(2-methylenecyclopropyl)alanine] as a natural product (see p. 129). Earlier, Burroughs (1960) had isolated 1-aminocyclopropyl-1-carboxylic acid (I) from *Pyrus communis* and the cowberry, *Vaccininum vitis-idaea*.

$$
\begin{array}{c}
\overset{\displaystyle H \diagdown \diagup H}{C} \\
\diagup \diagdown \\
H_2C\!\!-\!\!-\!\!C(NH_2)COOH
\end{array}
$$

1-Aminocyclopropyl-
1-carboxylic acid

(I)

While this compound shared the common group feature of the cyclopropyl ring, it was distinctive in being a saturated representative.

Fowden *et al.* (1969) subsequently obtained from *Aesculus parviflora* *exo*(*cis*)-3,4-methanoproline (II). Its structural features must have been unexpected, for this compound has an external methylene function. It is an important constituent of the nonprotein amino acids of the seed.

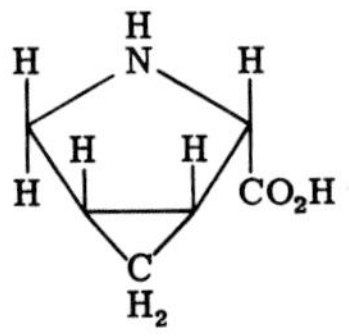

exo (*cis*) 3,4-
Methanoproline
(II)

At the same time, two geometric isomers having a carboxyl group in lieu of the external methylene group were characterized from *Aesculus parviflora* (Fowden *et al.*, 1969).

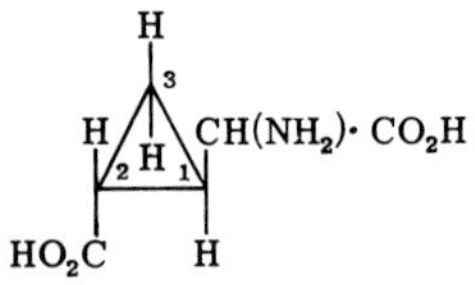

trans-2-(2-Carboxycyclo-   *cis*-2-(2-Carboxycyclo-
propyl)glycine   propyl)glycine
(III)   (IV)

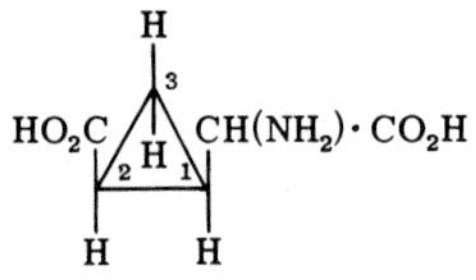

In contrast, *Blighia sapida* provided only the *trans* isomeride (Fowden *et al.*, 1970) but its 4-glutamyl peptide was also present (Fowden and Smith, 1968). An allied compound, *trans*-2-(2-carboxymethylcyclopropyl)glycine (V), is stored by *Blighia unijugata* (Fowden *et al.*, 1972a); a detectable level of the corresponding 4-glutamyl peptide was not found in this plant.

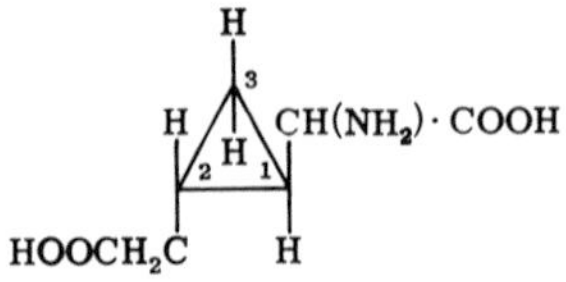

trans-2-(2-Carboxymethyl-
cyclopropyl)glycine
(V)

A final component in this series is the 3-methylated derivative of hypoglycin, i.e., 3-(methylenecyclopropyl)-3-methylalanine (VI) which is synthesized by *Aesculus californica* (Fowden and Smith, 1968).

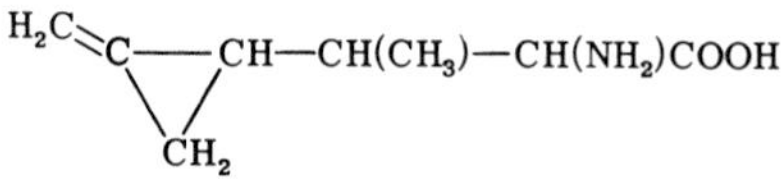

$$\underset{\text{CH}_2}{\overset{\text{H}_2\text{C}}{\diagdown}}\text{C}\text{---CH---CH(CH}_3)\text{---CH(NH}_2)\text{COOH}$$

3-(2-Methylenecyclo-
propyl)-3-methylalanine

(VI)

Fowden *et al.* (1969) took the above phytochemical findings to suggest a taxonomic linkage between *Aesculus* and *Blighia.* This point is also supported by the fact that *Aesculus californica* stores several C_7 amino acids such as 2-amino-4-methylhex-4-enoic acid (VII), 2-amino-4-methyl-6-hydroxyhex-4-enoic acid (VIII), and 2-amino-4-methylhexanoic acid (IX) while *Blighia unijugata* produces the isomeric 2-amino-5-methyl-6-hydroxyhex-4-enoic acid (X) (Fowden *et al.*, 1972a). Of the aforementioned *Aesculus* compounds, the first (VII) is a significant component of the seed soluble nitrogen resource as it represents nearly one-half of the total pool while its 4-glutamyl peptide along with the latter two compounds (IX and X) are minor components (Fowden and Smith, 1968).

$$\text{H}_3\text{C---CH}{=}\text{C(CH}_3)\text{---CH}_2\text{---CH(NH}_2)\text{COOH} \qquad \text{H}_3\text{C---CH}_2\text{---CH(CH}_3)\text{---CH}_2\text{---CH(NH}_2)\text{COOH}$$

(VII) (IX)

$$\text{HO---CH}_2\text{---CH}{=}\text{C(CH}_3)\text{CH}_2\text{---CH(NH}_2)\text{COOH} \qquad \text{HO---CH}_2\text{---C(CH}_3){=}\text{CH---CH}_2\text{---CH(NH}_2)\text{COOH}$$

(VIII) (X)

Additional unsaturated compounds, having the distinctive acetylenic bond, are known from study of the seeds of *Euphorbia longan* (Sung *et al.*, 1969). These include 2-amino-4-methylhex-5-ynoic acid (XII), 2-amino-4-hydroxymethylhex-5-ynoic acid (XIII), and 2-amino-4-hydroxyhept-6-ynoic acid (XIV); this plant also synthesizes an ethylenic compound, namely, 2-amino-4-hydroxyhept-6-enoic acid (XV).

$$\text{HC}{\equiv}\text{C---CH(CH}_3)\text{---CH}_2\text{---CH(NH}_2)\text{COOH} \qquad \text{HC}{\equiv}\text{C---C(CH}_2\text{OH)---CH}_2\text{---CH(NH}_2)\text{COOH}$$

(XII) (XIII)

$$\text{HC}{\equiv}\text{C---CH}_2\text{---CH(OH)---CH}_2\text{---CH(NH}_2)\text{COOH} \qquad \text{H}_2\text{C}{=}\text{CH---CH}_2\text{---CH(OH)---CH}_2\text{CH(NH}_2)\text{COOH}$$

(XIV) (XV)

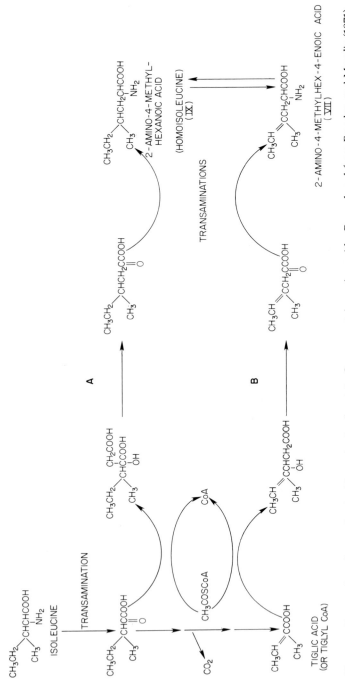

Fig. 45. Proposed pathways for the biosynthesis of certain C_7 nonprotein amino acids. Reproduced from Fowden and Mazelis (1971).

Metabolic Interrelationships

Fowden (1970) has integrated the metabolism of these structurally related compounds and suggested that 2-(methylenecyclopropyl)glycine could be an intermediate in the formation of the diastereoisomeric compounds III and IV. This would require an oxidation of the exocyclic carbon, perhaps with the production of a hydroxymethyl derivative between 2-(methylenecyclopropyl)glycine and II plus IV. Reductive cyclization of IV could produce *exo(cis)*-3,4-methanoproline (II); this reaction in *Aesculus parviflora* may bear reaction analogy with proline production from glutamic acid.

As part of the biosynthesis of certain C_7 amino acids, isoleucine chain elongation can occur by alteration of isoleucine by transamination and decarboxylation reactions to form tiglyl-CoA prior to addition of a two carbon fragment such as by condensation with acetyl-CoA (Boyle and Fowden, 1971). Eventual transamination of the addition product can lead to VII (Fig. 45, pathway **B**). Pathway (**A**) (Fig. 45) represents an alternative scheme in which the two carbon fragment addition occurs prior to decarboxylation and homoisoleucine (IX) is dehydrogenated to form VII. These schemes stand in contrast to the generation of a C_1-fragment from methionine or serine and its fusion with a terminal C_6-methyl group of leucine (Fowden and Mazelis, 1971). A third but far more remote biosynthetic possibility results from the isolation from the basidiomycete *Amanita solitaria* of 2-aminohexa-4,5-dienoic acid (Chilton *et al.*, 1968) which can be methylated to form 2-amino-4-methylhex-4-enoic acid (VII).

These possibilities were evaluated by assessing incorporation rates of appropriate ^{14}C-labeled precursor compounds. The transfer of C_1 groups, obtained from methionine or serine to either VII or IX, was invalidated by these precursor feeding studies. When [1-^{14}C]isoleucine was provided, no radioactivity appeared in VII in spite of the fact that [U-^{14}C]isoleucine labeled VII more effectively than any other tested compound. This finding is consistent with a loss of one carbon, possibly by decarboxylation, prior to the biosynthesis of 2-amino-4-methylhex-4-enoic acid (VII). Since the radioactive carbon of acetate moves into VII, it is reasonable to assume that the addition of a C_2 fragment to the C_5 chain could generate the C_7 carbon chain required for producing VII and IX. The fungal allenic compound was not found to be an effective precursor substance, thereby discounting this remote possibility from consideration. Since metabolism of [U-^{14}C]isoleucine resulted in the formation of VII and IX having essentially equal specific activity, and the seed concentration of VII is about 100 times greater than that of homoisoleucine (IX), the pathway designated **A** and analogous to valine transformation to leucine seemed highly unlikely. However, Fowden and Mazelis (1971) prudently noted that since the pools of these secondary metabolites are not necessarily in equilibrium, the lower pathway, i.e., **B** involving the critical formation of tiglic acid or its acyl-CoA derivative, cannot be unequivocally proposed as the physiological reaction mechanism.

In an ensuing follow-up study, an attempt was made to refine and distinguish between these reaction pathway alternatives by feeding later in the seed's ontogenetic development and by more effective precursor administration. In these ways, $[U\text{-}^{14}C]$isoleucine incorporation into 2-amino-4-methylhex-4-enoic acid was enhanced some 6-fold over earlier experiments. Cold tiglate (supplied in the NH_4^+ form) drastically reduced the incorporation of the labeled isoleucine into VII while increasing the radioactivity of the tiglate pool. Supplying unlabeled homoisoleucine, however, had little effect on ^{14}C incorporation into VII. In another relevant experiment, ^{14}C-labeled homoisoleucine and 2-amino-4-methylhex-4-enoic acid were supplied to the developing fruit. The latter compound was hydroxylated to form the 6-hydroxy derivative of VII and about 0.8% of the initially supplied radioactivity of VII moved into IX. However, no detectable conversion of ^{14}C-homoisoleucine into VII was reported. These experiments present strong evidence supporting the contention of Fowden and co-workers that pathway **B** represents the functional means for the biosynthesis of these C_7 amino acids (see also Boyle and Fowden, 1971).

G. MISCELLANEOUS COMPOUNDS

1. Albizziine

Albizzia julibrissin and *A. lophantha* were the initial biological sources for a novel ninhydrin-positive amino acid which also proved strongly responsive to Ehrlich's reagent (Gmelin *et al.*, 1958, 1959). Acid hydrolysis of this natural product yielded CO_2, NH_3, and 2,3-diaminopropionic acid which provided a meaningful structural clue; two isomerides were still possible.

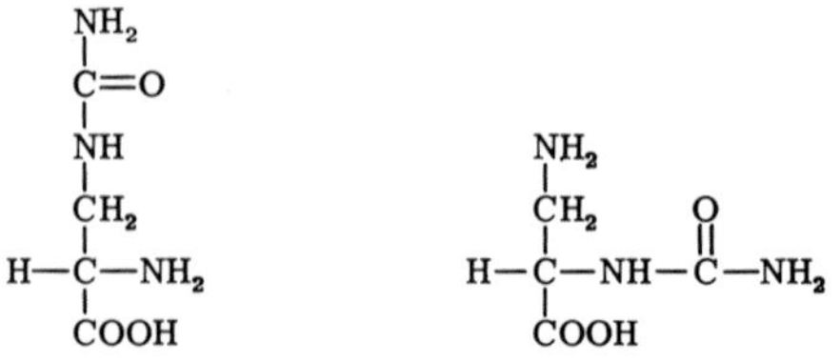

2-Amino-3-ureido-   2-Ureido-3-amino-
propionic acid   propionic acid

After synthesizing both isomerides, the identity of the natural product was gleaned by infrared spectral and rotatory data to be homologous to citrulline (Kjaer *et al.*, 1959). Quite recently, albizziine was isolated from four species of *Dislium*, a Sri Lankan endemic of the Caesalpiniaceae (Peiris and Seneviratne, 1977). These workers raised the speculation of a phylogenetic link between the Mimosaceae and certain Caesalpiniceae due to their concurrent production of this ureido-containing amino acid. While this may prove valid, such conjecture based only on a single line of evidence is tenuous at best, often wrong, and ought to be avoided.

Albizziine is a common component of many *Acacia* and *Albizzia* species (Seneviratne and Fowden, 1968a) where it often constitutes the principal component of the free amino acid fraction of the seed. Its structural analogy with glutamine suggests an appreciable potential for acting as a competitive antagonist of the protein amino acid, particularly in blocking vital nitrogen transfer reactions.

$$
\begin{array}{cc}
\underset{|}{NH_2} & \underset{|}{NH_2} \\
\underset{|}{C=O} & \underset{|}{C=O} \\
\underset{|}{NH} & \underset{|}{CH_2} \\
\underset{|}{CH_2} & \underset{|}{CH_2} \\
H-\underset{|}{C}-NH_2 & H-\underset{|}{C}-NH_2 \\
COOH & COOH \\
\text{L-Albizziine} & \text{L-Glutamine}
\end{array}
$$

This point has been tested experimentally by the biochemists at Rothamsted Experimental Station in England with asparagine synthetase and several glutamate synthases. In *Lupinus albus*, an equimolar ratio of albizziine to glutamine strongly inhibits (over 70% curtailment of activity) the ability of the protein amino acid to act as an amide nitrogen donor to aspartic acid (Lea and Fowden, 1975b). It is much more usual, however, for albizziine to be provided in a molar excess relative to glutamine in order to obtain significant curtailment of various reactions involved in glutamine-dependent amide transfer (Table 16).

Moreover, albizziine does not compete effectively with glutamine for the active site of glutaminyl-tRNA synthetase in non-producing organisms and more than a 200-fold excess of this nonprotein amino acid is required for 50% inhibition of glutamine activation (Lea and Fowden, 1973). No significant difference was reported in the behavior of this enzyme from *Phaseolus aureus* (nonproducing species) and *A. julibrissin*. This inability of albizziine to cause aberrant protein formation and its relative impotency as a glutamine antagonist would be expected to markedly decrease its overall antimetabolic activity. This point has been evaluated by biological testing of this natural product in Rosenthal's laboratory with the insect *Manduca sexta* as well as *Lemna minor*. This testing and comparable biological evaluations against the bruchid beetle, *Callosobruchus maculatus*, and *Lactuca sativa* (lettuce) sustained the expectation that this nonprotein amino acid would lack significant biological toxicity (see p. 141).

In a perceptive and stimulating presentation, Doyle McKey (1979) analyzed the cogent arguments for a selective advantage to the producing organism of utilizing a particular plant metabolite concurrently for nitrogen storage as well as chemical defense. In this way, some component of the defensive investment can be retrieved by diverting stored molecular energy into primary metabolic pathways. The evidence for such a duality in function for canavanine is reaching the level of incontrovertibility—but what about albizziine?

Both albizziine and canavanine are important nitrogen-storing metabolites of particular leguminous species and both are known to function as an antagonist of a protein amino acid although albizziine is much less potent than canavanine. Yet, biological testing of albizziine provided no evidence for such a duality in function as mentioned above. In 1978, Navon and Bernays reported on their evaluation of the action of albizziine with two insects, namely, *Locusta* and *Chortoicetes*. Experiments with these graminivorous feeders culminated in the first report of albizziine's insecticidal properties. In fact, it is one of the most toxic of the nonprotein amino acids tested in this system—exceeding even canavanine. This study is important for its demonstration of the an-

Table 16 Albizziine as an Inhibitor of Certain Glutamine-Dependent Amide Transferases of Higher Plants[a]

Enzyme	Source	Concentration (mM)		Inhibition (%)
		Glutamine	Albizziine	
Asparagine synthetase	*Lupinus* seedlings	1.0	5.0	85.2
Glutamate synthase (NADH)	*Pisum* roots	5.0	10.0	89
Glutamate synthase (ferredoxin)	*Vicia* leaves	2.5	10.0	45
Glutamate synthase (NADH)	*Phaseolus* root nodules	5.0	20.0	83.4

[a] Reproduced with permission from Fowden and Lea (1979).

timetabolic properties of albizziine and its support of the contention that many, but of course not all, nitrogen-storing nonprotein amino acids can also function in plant defense. This work also emphasizes that the nature of our concept of the toxicity of a particular secondary plant metabolite is relative to how and where one evaluates its toxicity.

Metabolism

Essentially nothing is known of albizziine metabolism. It has been suggested that albizziine may be biosynthesized by the reaction of O-acetylserine and urea but no precedence exists for attributing such a nitrogen-donating role to urea. The alternative suggestion of an addition reaction to 2,3-diaminopropionic acid is much more viable (Fowden *et al.*, 1979). The required ureido group can be provided by carbamoyl phosphate; such a carbamylation reaction would parallel both citrulline and ureidohomoserine formation. It is prudent to avoid repetitious postulation of a family of biosynthetic reactions sharing analogy to the ornithine–urea cycle for every compound bearing structural analogy to the constituents of this cycle. On the other hand, there is no reason to assume that albizziine cannot participate in a sequential reaction series involving such intermediates as 2,3-diaminopropionic acid and 2-amino-3-guanidinopropionic acid (a lower homolog of ornithine and arginine, respectively). It is mentioned only because if shown to be so, it would certainly represent a fascinating example of multiple higher plant proliferation of certain structurally similar nonprotein amino acids.

2. 4-Glutamyl Dipeptides

Meister (1973) and associates have proposed a cyclic reaction sequence that is responsible for amino acid reabsorption in the kidney. In this reputed process, glutathione reacts with a given amino acid to form the corresponding 4-glutamyl dipeptide and cysteinylglycine. A 4-glutamylcyclotransferase liberates the transported amino acid and 5-oxoproline. The latter is phosphorylated to regenerate the glutamic acid needed to resynthesize glutathione.

Higher plants are known to synthesize a number of 4-glutamyl dipeptides that involve covalent peptide bond formation between the γ-COOH group of glutamic acid and a full range of nonprotein amino acid structures (Table 17). γ-Glutamyl transpeptidase, the catalytic protein forming these natural products, has been isolated from the legume *Phaseolus vulgaris*; glutathione was suggested as the physiological donor of the glutamyl group (Goore and Thompson, 1967).

Glutathione + amino acid → 4-glutamyldipeptide + cysteinylglycine

Moreover, 5-oxoprolinase is widely distributed in higher plants where it catalyzes the following reaction (Mazelis and Creveling, 1978):

$$\text{5-Oxo-L-proline} + \text{ATP} + H_2O \underset{K^+ \text{ or } NH_4^+}{\overset{Mg^{2+}}{\rightleftharpoons}} \text{L-glutamic acid} + \text{ADP} + P_i$$

Table 17   The 4-Glutamyl Dipeptides of Higher Plants[a]

Nonprotein amino acid moiety	Sources
S-Methylcysteine	*Phaseolus, Astragalus, Allium*
S-Methylcysteine sulfoxide	*Phaseolus, Astragalus, Allium*
S-(2-Carboxypropyl)cysteine	*Allium*
S-(Propyl)cysteine	*Allium*
S-(Prop-1-enyl)cysteine	*Santalum, Allium*
S-(Prop-1-enyl)cysteine sulfoxide	Legumes, *Allium*
S-Allylcysteine	*Allium*
S-Allylmercaptocysteine	*Allium*
N,N'-bis(4-glutamyl)-3,3'-(1-methyl-ethylene-1,2-dithio)alanine	*Allium*
Djenkolic acid	*Acacia*
Djenkolic acid sulfoxide	*Acacia*
Cystathionine	*Astragalus*
3-(Pyrazol-1-yl)alanine	Curcubitaceae
Willardiine	*Fagus*
Pipecolic acid	*Gleditsia*
Lathyrine	*Lathyrus*
Albizziine	*Acacia*
2-Amino-4-methylhex-4-enoic acid	*Aesculus*
2-Amino-5-methyl-6-hydroxyhex-4-enoic acid	*Aesculus*
3-Cyanoalanine	*Lathyrus, Vicia*
3-Alanine	*Lunaria, Iris*
Homoserine	Ubiquitous
3-Aminoisobutyric acid	*Iris*
4-Aminobutyric acid	*Lunaria, Vicia, Beta*
4-Aminoisobutyric acid	*Iris, Lunaria*
Se-methylselenocysteine	*Astragalus*
Selenocystathionine	*Astragalus*
3-(Methylenecyclopropyl)alanine	*Blighia, Billia, Acer*
2-(Methylenecyclopropyl)glycine	*Billia, Litchi, Acer*
2-(Carboxycyclopropyl)glycine	*Blighia*
Theanine	*Thea*
2-Amino-3-methylenepentanoic acid	*Philadelphus*
2-Amino-3-methylene-4-pentenoic acid	*Philadelphus*
3-(Isoxazolin-5-one-2-yl)alanine	*Pisum, Lathyrus*

[a] For appropriate references see: Waley (1966), Synge (1968), Thompson *et al.* (1964), Steward and Durzan (1965), and Kasai and Larsen (1980).

Recently, Kean and Hare (1980) isolated a 4-glutamyl transpeptidase from the immature fruit of the ackee, *Blighia sapida,* that uses glutathione as the glutamyl donor. In this plant, 3-(methylenecyclopropyl)alanine (hypoglycin A) yields the glutamyl dipeptide known trivially as hypoglycin B (see also p. 128). While it is evident that higher plants synthesize many dipeptides, the *raison d'être* for their production and storage is not clear. It may be that amino acid transportation is somehow facilitated as the 4-glutamyl dipeptide. No experimental evidence has been obtained for a γ- or 4-glutamyl cycle in higher plants analogous to the cyclic reaction sequence proposed for mammals.

Some plants, e.g., members of the genus *Allium* or *Iris,* which are known to be rich sources of these compounds, do not appear to have much or any discernible transpeptidase activity. This suggests that this enzyme may not be the only mechanism by which plants produce these dipeptides; information on alternative pathways is not available. It is evident, however, that it is not easy to rationalize the storage of a nonprotein amino acid as its 4-glutamyl dipeptide as compared to the free compound especially when proteins and energy are required both for the complexing and eventual release of the nonprotein amino acid moiety.

3. 4-Aminobutyric Acid and 3-Alanine

4-Aminobutyric acid, first isolated by Okunuki (1937), is a major constituent of higher plants. Although its overall physiological function in the plant is only poorly understood, its accumulation is enhanced under adverse environmental conditions (Inatomi and Slaughter, 1970). The ubiquitous distribution of L-glutamic acid decarboxylase which produces this compound probably accounts for its essentially universal occurrence. The amino group functions in amino acid transamination and thus it supplies nitrogen for other plant needs. Additionally, the resulting semialdehyde can be oxidized to succinic acid which provides another interchange between amino acid metabolism and the reactions of the Krebs cycle. The use of ^{14}C-labeled substrates and carrot explant cultures permitted Steward *et al.* (1956) to postulate that 4-aminobutyric acid might be a precursor for glutamic acid and eventually glutamine. The importance of this reversal in the usually conceived route of 4-aminobutyric acid-glutamate interchange has not been studied adequately. Recently, Terano and Suzuki (1978) reported the biosynthesis of 4-aminobutyric acid from spermine in *Zea mays.* Since this reaction was thought originally to be limited to prokaryotes, the observations of Terano and Suzuki open the possibility that other reactions of microorganisms involving this compound might also be manifested by higher plants.

Tixier and Desmaison (1980) analyzed the free amino acids of the fruit of *Castanea sativa* and reported high levels of 4-aminobutyric acid. This was taken to result from the appreciable arginine in the seed which provided ornithine. The latter compound was converted to glutamic acid via glutamic acid semialdehyde prior to the decarboxylation of the protein amino acid.

Decarboxylation of L-aspartic acid yields 3-alanine which condenses with pantoic acid to give pantothenic acid, a known precursor of coenzyme A. This biosynthetic reaction is well established in microorganisms but is wholly putative for higher plants. 3-Alanine is also a component of 4-glutamylcysteine-3-alanine or homoglutathione in which 3-alanine replaces the alanine moiety of glutathione. In mung bean, homoglutathione occurs in much higher levels than glutathione itself and probably represents a primary free thiol compound of this legume.

ADDITIONAL READING

Beevers, L. (1970). "Nitrogen Metabolism in Plants." Am. Elsevier, New York.

Bidwell, R. G. S., and Durzan, D. J. (1975). Some recent aspects of nitrogen metabolism. *In* "Historical and Current Aspects of Plant Physiology" (P. J. Davies, ed.), pp. 152–225. New York State College of Agriculture and Life Sciences, Ithaca, New York.

Meister, A. (1965). "Biochemistry of the Amino Acids," 2nd ed., 2 vols. Academic Press, New York.

Miflin, B. J., ed. (1980). "The Biochemistry of Plants," Vol. 5. Academic Press, New York.

Miflin, B. J., and Lea, P. J. (1977). "Amino Acid Metabolism" *Annu. Rev. Plant Physiol.* **28,** 95–120.

Steward, F. C., and Durzan, D. J. (1965). Metabolism of nitrogenous compounds. *In* "Plant Physiology: A Treatise" (F. C. Steward, ed.), Vol. 4A, pp. 664–679. Academic Press, New York.

SUMMARY OF APPENDIX

Group No.	Group*	Compound No.
I	Monoaminomonocarboxylic acids	1–12c
II	Monoaminodicarboxylic acids and their amide derivatives	13–25
III	Diaminomonocarboxylic acids and their derivatives	26–40
IV	Diaminodicarboxylic acids	41
V	Hydroxy and hydroxy-substituted acids	41b–100
VI	Ureido and guanidino acids and similar derivatives	101–111
VII	Aromatic constituents	112–130
VIII	Heterocyclic compounds	131–170
IX	N-substituted derivatives	171–184
X	Sulfur-containing compounds	185–214c
XI	Selenium-containing compounds	215–223
XII	Cyclopropyl ring-containing compounds	224–229

* In organizing the nomenclature of these natural products, names of long standing, e.g., levulinic, adipic, or pimelic acid, were retained. Otherwise, the aliphatic chain was named from the appropriate organic acid root: propionic, butyric, pentanoic, hexanoic, or heptanoic. Only numerical designations were employed to locate substituents on the chain, and the substituent groups were taken in numerical sequence.

APPENDIX

NONPROTEIN AMINO AND IMINO ACIDS AND THEIR DERIVATIVES*

No.	Compound	Source	Reference
Group I: Monoaminomonocarboxylic acids			
1	3-Alanine	*Iris tingitana*	*136*
	$H_2N—CH_2—CH_2—CO_2H$	Various plants	*125*
2	3-Cyanoalanine	*Vicia sativa*	*158*
	$N≡C—CH_2—CH(NH_2)CO_2H$		
3	2-Aminobutyric acid	*Pisum*	*208*
	$H_3C—CH_2—CH(NH_2)CO_2H$		
4	4-Aminobutyric acid	Widely distributed	*174*
	(piperidinic acid)		
	$H_2N—(CH_2)_3—CO_2H$		
5	2-Methylene-4-aminobutyric acid	*Arachis hypogaea*	*56*
	$H_2N—CH_2—CH_2—C(=CH_2)—CO_2H$		
6	3-Methylene-4-aminobutyric acid	*Tulipa*	*57*
	$H_2N—CH_2—C(=CH_2)—CH_2—CO_2H$		

*In organizing the arrangement of these natural products, (1) hydroxylated compounds were placed in group V except for phenyl aromatic members which constitute group VII and the N-substituted derivatives of group IX, and (2) N-substituted derivatives comprise group IX except when the substituted nitrogen atom is committed to ring structure formation. In such cases, it is organized with the heterocyclic compounds of group VIII. See Table 17 for the γ- or 4-glutamyl dipeptides of the nonprotein amino acids. This compilation includes all green plants and variously colored algae. Many putative nonprotein amino acid intermediates of plant metabolism were not included, although they undoubtedly exist, since they have not yet been isolated from appropriate sources.

(*continued*)

APPENDIX (*Continued*)

No.	Compound	Source	Reference
7	2-Aminoisobutyric acid $(CH_3)_2$—$C(NH_2)CO_2H$	*Iris tingitana*	35
8	3-Aminoisobutyric acid H_2N—CH_2—$CH(CH_3)CO_2H$	*Iris tingitana*	3
9	5-Aminolevulinic acid H_2N—CH_2—$C(=O)$—CH_2—CH_2—CO_2H	Widespread occurrence	
10	2-Amino-4-methylhexanoic acid (homoisoleucine) 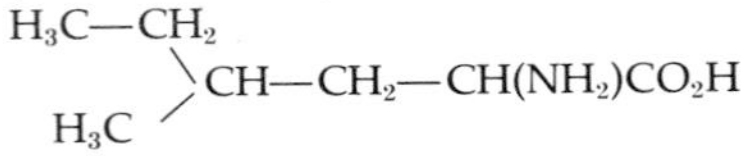 H_3C—CH_2 H_3C ⟩CH—CH_2—$CH(NH_2)CO_2H$	*Aesculus californica*	58
11	2-Amino-4-methylhex-4-enoic acid 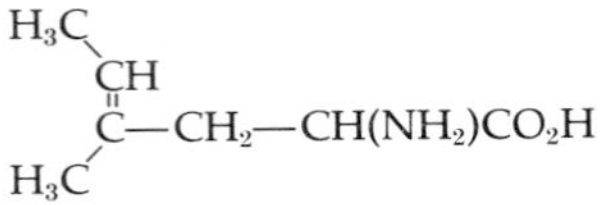 H_3C CH C—CH_2—$CH(NH_2)CO_2H$ H_3C	*Aesculus californica*	58
12a	2-Amino-4-methylhex-5-ynoic acid $HC{\equiv}C$—$CH(CH_3)$—CH_2—$CH(NH_2)CO_2H$	*Euphorbia longan*	178
12b	2-Amino-3-methylenepentanoic acid 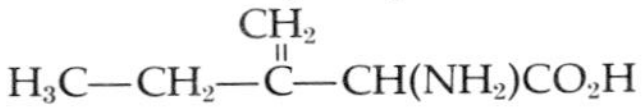 CH_2 $\parallel$ H_3C—CH_2—C—$CH(NH_2)CO_2H$	*Philadelphus coronarius*	206
12c	2-Amino-3-methylene-4-pentenoic acid 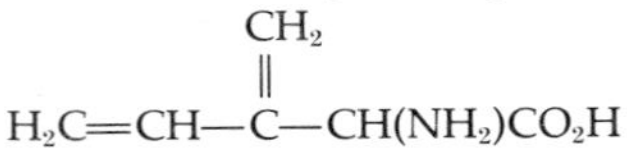 CH_2 $\parallel$ $H_2C{=}CH$—CH—C—$CH(NH_2)CO_2H$	*Philadelphus coronarius*	206

Group II: Monoaminodicarboxylic acids and their amide derivatives

No.	Name / Structure	Source	Ref.
13	2-Aminoadipic acid $H_2OC—(CH_2)_3—CH(NH_2)CO_2H$	*Pisum sativum* *Zea mays*	*82* *218*
14	3-Aminoglutaric acid $H_2OC—CH_2—CH(NH_2)—CH_2—CO_2H$	*Chondria armata*	*183*
15	2-Aminopimelic acid $H_2OC—(CH_2)_4—CH(NH_2)CO_2H$	*Asplenium septentrionale*	*198*
16	N^4-Ethylasparagine $H_3C—CH_2—NH—C(=O)—CH_2—CH(NH_2)CO_2H$	*Tulipa gesneriana* *Ecballium elaterium* *Bryonia dioica*	*51* *74* *49*
17	N^4-Methylasparagine $H_3C—NH—C(=O)—CH_2—CH(NH_2)CO_2H$	*Corallocarpus epigaeus*	*34*
18	4-Ethylideneglutamic acid $H_3C—CH$ $\ \ \ \ \ \|$ $H_2OC—C—CH_2—CH(NH_2)CO_2H$	*Tulipa* *Mimosa* *Tetrapleura tetraptera*	*51* *65* *65*
19a	4-(β-D-galactopyranosyloxy)-4-isobutylglutamic acid $\ \ \ \ \ \ \ \ \ O—gal$ $\ \ \ \ \ \ \ \ \ \|$ $H_2OC—C—CH_2—CH(NH_2)CO_2H$ $\ \ \ \ \ \ \ \ \ \|$ $\ \ \ \ \ \ \ CH_2—CH(CH_3)_2$	(2S, 4R): *Reseda odorata*	*118*
19b	2,4-Methanoglutamic acid (1-amino-1,3-dicarboxycy-clobutane) 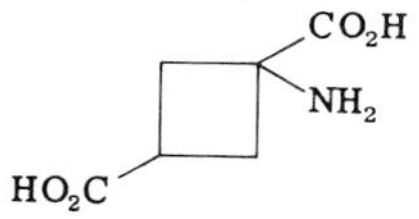 	*Ateleia herbert smithii*	*12b*
20	*erythro*-4-Methylglutamic acid $H_2OC—CH(CH_3)—CH_2—CH(NH_2)CO_2H$	*Phyllitis scolopendrium* *Tulipa gesneriana*	*199* *13*

(continued)

APPENDIX (*Continued*)

No.	Compound	Source	Reference
21	4-Methyleneglutamic acid $H_2OC-C(=CH_2)-CH_2-CH(NH_2)CO_2H$	*Arachis hypogaea* *Amorpha fruticosa* *Tulipa gesneriana* *Tetrapleura tetraptera* *Lunaria annua*	33 219 219 65 115
22	N^5-Isopropylglutamine $HN-C(=O)-CH_2-CH_2-CH(NH_2)CO_2H$ ⎮ HC ／＼ H_3C CH_3	*Thea*	108
23	N^5-Methylglutamine $H_3C-NH-C(=O)-CH_2-CH_2-CH(NH_2)CO_2H$	*Arachis hypogaea* *Amorpha fruticosa* *Tulipa gesneriana* *Phyllitis scolopendrium* *Thea*	33 219 193 13 163
24	4-Methyleneglutamine $H_2N-C(=O)-C(=CH_2)-CH_2-CH(NH_2)CO_2H$		
25	Theanine N^5-Ethylglutamine $H_3C-CH_2-NH-C(=O)-CH_2-CH_2-CH(NH_2)CO_2H$		

Group III: Diaminomonocarboxylic acids and their derivatives

No.	Compound	Source	Reference
26	Canaline [2-amino-4-(aminooxy)butyric acid] $H_2N-O-CH_2-CH_2-CH(NH_2)CO_2H$	*Canavalia ensiformis* *Astragalus sinicus*	132 92
27	2,4-Diaminobutyric acid $H_2N-CH_2-CH_2-CH(NH_2)CO_2H$	*Acacia* *Lathyrus latifolius* *Polygonatum multiflorum* *Arion empiricorum*	42 159 55 1

28	N^4-Acetyl-2,4-diaminobutyric acid $H_3C—C(=O)—NH—CH_2—CH_2—CH(NH_2)CO_2H$	Euphorbia pulcherrima (latex)	122
29	N^4-Lactyl-2,4-diaminobutyric acid $H_3C—CH(OH)—C=O$ $\backslash$ $HN—CH_2—CH_2—CH(NH_2)CO_2H$	Beta vulgaris Beta vulgaris	53 53
30	N^4-Oxalyl-2,4-diaminobutyric acid $H_2OC—C(=O)—NH—CH_2—CH_2—CH(NH_2)CO_2H$	Acacia Lathyrus latifolius	42 8
31	2,3-Diaminopropionic acid $H_2N—CH_2—CH(NH_2)CO_2H$	Mimosa palmeri Acacia	71 42
32	N^3-Acetyl-2,3-diaminopropionic acid $H_3C—C(=O)—NH—CH_2—CH(NH_2)CO_2H$	Acacia	168
33	N^3-Methyl-2,3-diaminopropionic acid $H_3C—NH—CH_2—CH(NH_2)CO_2H$	Cycas circinalis	197
34	N^3-Oxalyl-2,3-diaminopropionic acid $H_2OC—C(=O)—NH—CH_2—CH(NH_2)CO_2H$	Lathyrus sativus	154b
35	N^6-Acetyllysine $H_3C—C(=O)—NH—(CH_2)_4—CH(NH_2)CO_2H$	Beta vulgaris	53
36	N^6-Methyllysine $H_3C—NH—(CH_2)_4—CH(NH_2)CO_2H$	Sedum acre Sedum sarmentasum	119
37	N^6-Trimethyllysine (laminine) $(CH_3)_3—\overset{+}{N}—(CH_2)_4—CH(NH_2)CO_2H$	Reseda luteola Laminaria	117 184
38	Ornithine (2,5-diaminopentanoic acid) $H_2N—(CH_2)_3—CH(NH_2)CO_2H$	Widely distributed	
39	N^5-Acetylornithine $H_3C—C(=O)—NH—(CH_2)_3—CH(NH_2)CO_2H$	Widespread occurrence	160

(continued)

APPENDIX (Continued)

No.	Compound	Source	Reference
40	Saccharopine [N^6-(2'-glutamyl)]lysine] $\begin{array}{c}\text{CO}_2\text{H}\\ \mid \\ \text{HC—NH—(CH}_2\text{)}_4\text{—CH(NH}_2\text{)CO}_2\text{H}\\ \mid \\ \text{(CH}_2\text{)}_2\\ \mid \\ \text{CO}_2\text{H}\end{array}$	(2S, 2'S): *Reseda odorata* may be widely distributed	173b 141b
Group IV: Diaminodicarboxylic acids			
41	2,6-Diaminopimelic acid $\text{HO}_2\text{C—CH(NH}_2\text{)—(CH}_2\text{)}_3\text{—CH(NH}_2\text{)CO}_2\text{H}$	*Pinus* (pollen)	25
Group V: Hydroxy and hydroxy-substituted acids			
41b	2-Amino-4-carboxy-4-hydroxyadipic acid $\begin{array}{c}\text{CO}_2\text{H}\\ \mid \\ \text{H}_2\text{OC—CH}_2\text{—C(OH)—CH}_2\text{—CH(NH}_2\text{)CO}_2\text{H}\end{array}$	*Caylusea abyssinica*	145b
42	4-Hydroxyarginine $\text{H}_2\text{N—C(=NH)—NH—CH}_2\text{—CH(OH)—CH}_2\text{—}$ $\text{CH(NH}_2\text{)CO}_2\text{H}$	*Vicia* spp.	9
43	N^4-(2-Hydroxyethyl)asparagine $\text{HO—CH}_2\text{—CH}_2\text{—NH—C(=O)—CH}_2\text{—CH(NH}_2\text{)—}$ CO_2H	*Bryonia dioica* *Ecballium elaterium*	49 74
44	*erythro*-3-Hydroxyaspartic acid $\text{H}_2\text{OC—CH(OH)—CH(NH}_2\text{)CO}_2\text{H}$	*Trifolium pratense* *Astragalus sinicus*	217 93

45	4-Hydroxycitrulline (2-Amino-4-hydroxy-5-ureidopentanoic acid) $H_2N-C(=O)-NH-CH_2-CH(OH)-CH_2-$ $CH(NH_2)CO_2H$	*Vicia faba* *Vicia fulgens* *Vicia unijuga*	94 11 11
46	S-(2-Hydroxy-2-carboxyethanethiomethyl)cysteine $HO_2C-CH(OH)-CH_2-S-CH_2-S-CH_2-$ $CH(NH_2)CO_2H$	*Acacia georginae*	95
47	Dichrostachinic acid [S-(3-hydroxy-3-carboxyethane sulfonyl methyl) cysteine] $HO_2C-CH(OH)-CH_2-SO_2-CH_2-S-$ $CH_2-CH(NH_2)CO_2H$	*Mimosa* *Dichrostachys glomerata* *Neptunia oleracea*	63 63 63
48	Fusarinine $HO-(CH_2)_2-C(CH_3)=CH-C(=O)-$ $N(OH)-(CH_2)_3-CH(NH_2)CO_2H$	Algae	41
49	*threo*-4-Hydroxyglutamic acid $H_2OC-CH(OH)-CH_2-CH(NH_2)CO_2H$	*Phlox decussata* *Hemerocallis fulva* *Tulipa gerheriana* (2S, 4R): *Phlox*	201 169 13 11b
50	3,4-Dihydroxyglutamic acid $H_2OC-CH(OH)-CH(OH)-CH(NH_2)CO_2H$	*Lepidium sativum* *Rheum rhaponticum*	200
51	3-Hydroxy-4-methylglutamic acid $H_2OC-CH(CH_3)-CH(OH)-CH(NH_2)CO_2H$	*Gleditsia amaphoides* (2S, 3S, 4R): *Gleditsia tria-* *canthos* (2S, 3R, 4S): *Gymnocladus* *dioicus* *Caesalpinia bonduc* *Caesalpinia major* *Gleditsia caspica*	215 215 27 215 215 118
52	3-Hydroxy-4-methyleneglutamic acid $H_2OC-C(=CH_2)-CH(OH)-CH(NH_2)CO_2H$		

(*continued*)

APPENDIX (*Continued*)

No.	Compound	Source	Reference

53	4-Hydroxy-4-methylglutamic acid
	H_2OC—$C((CH_3)$—CH_2—$CH(NH_2)CO_2H$
	       $\|$
	      OH
		(2S, 4S): *Pandanus veitchii*
		(2S, 4R): *Ledenbergia roseo-*
		aenea
		Adiantum pedatum
		(2S, 4S): *Phyllitis scolo-*
		pendrium
			99
			99
			
			79
			199

54	4-Hydroxyglutamine H_2N—$C(=O)$—$CH(OH)$—CH_2—$CH(NH_2)CO_2H$	*Phlox decussata* *Hemerocallis fulva*	*14* *47*
55	N^5-(2-Hydroxymethylbutadienyl)-*threo*-4-hydroxyglu- tamine (pinnatanine) H_2C=CH—$C((CH_2OH)$=CH—NH—$C(=O)$— $CH(OH)$—CH_2—$CH(NH_2)CO_2H$	*Staphylea pinnata*	*81*
56	N^5-(2-Hydroxyethyl)glutamine HO—$(CH_2)_2$—NH—$C(=O)$—$(CH_2)_2$—$CH(NH_2)$— CO_2H	*Lunaria annua*	*116*
57	2-Amino-4-hydroxyhept-6-enoic acid H_2C=CH—CH_2—$CH(OH)$—CH_2—$CH(NH_2)CO_2H$	*Euphorbia longan*	*178*
58	2-Amino-4-hydroxyhept-6-ynoic acid HC≡C—CH_2—$CH(OH)$—CH_2—$CH(NH_2)CO_2H$	*Euphorbia longan*	*178*
59	5-Hydroxynorleucine H_3C—$CH(OH)$—$(CH_2)_2$—$CH(NH_2)CO_2H$	*Crotalaria juncea*	*150*

60	2-Amino-4-methyl-6-hydroxylhex-4-enoic acid
	HO—CH_2—CH=C—CH_2—$CH(NH_2)CO_2H$
	             $\|$
	           CH_3
		Aesculus californica	*58*

61	2-Amino-5-methyl-6-hydroxyhex-4-enoic acid $HO-CH_2-C=CH-CH_2-CH(NH_2)CO_2H$ $\qquad\qquad\ \ \|$ $\qquad\qquad\ CH_3$	*Blighia unijugata*	61
62	2-Amino-4-hydroxymethylhex-5-ynoic acid $HC\equiv C-C(CH_2OH)-CH_2-CH(NH_2)CO_2H$	*Euphorbia longan*	178
63	*threo*-4-Hydroxyhomoarginine $H_2N-C(=NH)-NH-CH_2-CH_2-CH(OH)-$ $CH_2-CH(NH_2)CO_2H$	*Lathyrus*	6
64	S-Hydroxymethylhomocysteine $CH_2OH-S-CH_2-CH_2-CH(NH_2)CO_2H$	*Chondrus ocellatus*	181
65	Homoserine $HO-CH_2-CH_2-CH(NH_2)CO_2H$	*Pisum sativum*	78
66	O-Acetylhomoserine $H_3C-C(=O)-O-CH_2-CH_2-CH(NH_2)CO_2H$	*Pisum sativum*	78
67	O-Oxalylhomoserine $H_2OC-C(=O)-O-CH_2-CH_2-CH(NH_2)CO_2H$	*Lathyrus*	153
68	O-Phosphohomoserine $H_2O_3P-O-(CH_2)_2-CH(NH_2)CO_2H$	*Zea mays* Certain algae	107
69	4-Hydroxyisoleucine $H_3C-CH(OH)$ $\qquad\qquad\ \ \|$ $\qquad\qquad\ CH-CH(NH_2)CO_2H$ $\qquad\qquad\ \ \|$ $\qquad\qquad\ H_3C$	*Trigonella foenum-graecum*	59

(continued)

APPENDIX (*Continued*)

No.	Compound	Source	Reference
70	1-Carboxy-6,7-dimethoxy-8-hydroxy-1,2,3,4-tetrahydroisoquinoline 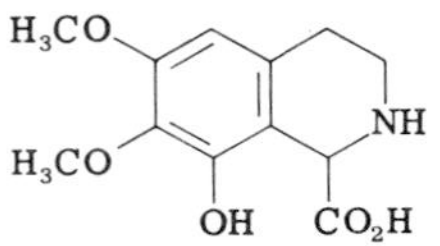	Peyote cactus	*102b*
71	3-Carboxy-6,7-dihydroxy-1,2,3,4-tetrahydroiso-quinoline 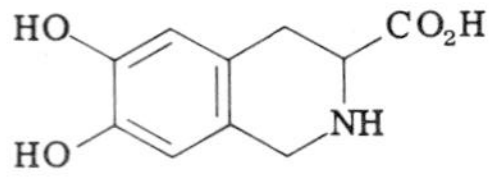	*Mucuna mutisiana*	*12*
72	1-Methyl-3-carboxy-6-hydroxy-1,2,3,4-tetrahydroisoquinoline 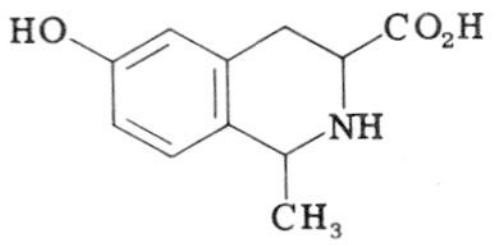	*Euphorbia myrsinites*	*138*
73	1-Methyl-3-carboxy-6,7-dihydroxy-1,2,3,4-tetrahydroisoquinoline 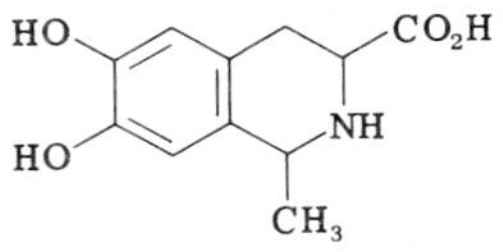	*Mucuna deeringiana*	*31*

74	*threo*-3-Hydroxyleucine 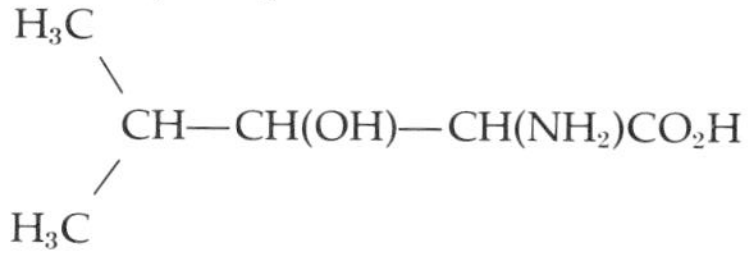 H_3C $\quad\diagdown$ $\quad\quad CH\!-\!CH(OH)\!-\!CH(NH_2)CO_2H$ $\quad\diagup$ H_3C	*Deutzia gracilis*	98
75	5-Hydroxyleucine 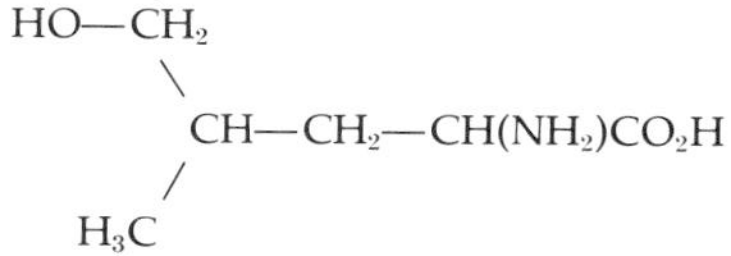 $HO\!-\!CH_2$ $\quad\quad\diagdown$ $\quad\quad\quad CH\!-\!CH_2\!-\!CH(NH_2)CO_2H$ $\quad\quad\diagup$ $\quad H_3C$	*Deutzia gracilis*	97
76	2-Hydroxylysine 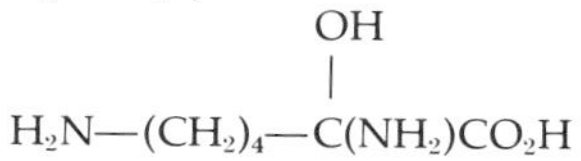 $\quad\quad\quad OH$ $\quad\quad\quad\,\mid$ $H_2N\!-\!(CH_2)_4\!-\!C(NH_2)CO_2H$	*Salvia officinalis*	16
77	4-Hydroxylysine $H_2N\!-\!(CH_2)_2\!-\!CH(OH)\!-\!CH_2\!-\!CH(NH_2)CO_2H$	*Salvia officinalis*	16
78	5-Hydroxylysine $H_2N\!-\!CH_2\!-\!CH(OH)\!-\!(CH_2)_2\!-\!CH(NH_2)CO_2H$	*Medicago*	217
79	N^6-Acetyl-5-hydroxylysine $H_3C\!-\!C(\!=\!O)\!-\!NH\!-\!CH_2\!-\!CH(OH)\!-\!CH_2\!-$ $CH_2\!-\!CH(NH_2)CO_2H$	*Beta vulgaris*	53
80	N^6-Trimethyl-5-hydroxylysine $\overset{+}{(CH_3)_3}\!-\!N\!-\!CH_2\!-\!CH(OH)\!-\!CH_2\!-\!CH_2\!-$ $CH(NH_2)CO_2H$	Diatoms	141

(continued)

215

APPENDIX (*Continued*)

No.	Compound	Source	Reference
81	Mimosine [3-(*N*-3-hydroxy-4-pyridone)—2-aminopropionic acid] 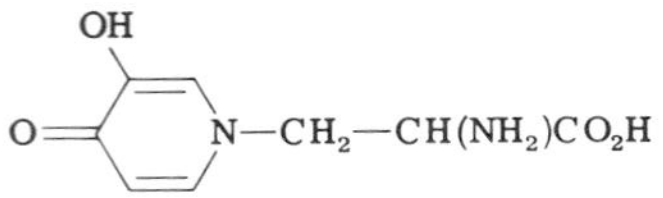 $N—CH_2—CH(NH_2)CO_2H$	*Mimosa pudica* *Leucaena leucocephala*	*157* *15*
82	Mugineic acid N-[3-(3-hydroxy-3-carboxypropylamino)-2-hydroxy-3-carboxypropyl]-azetidine-2-carboxylic acid 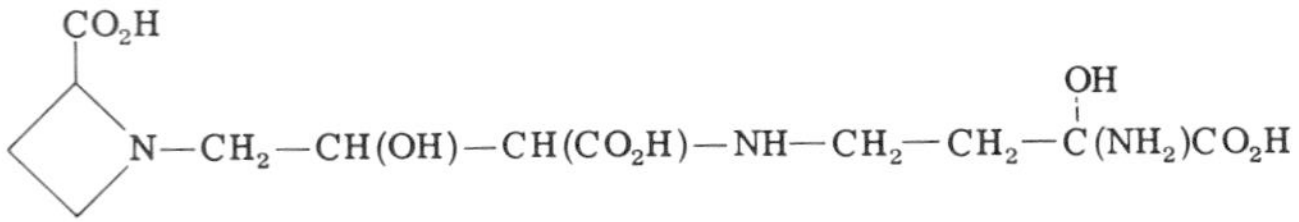 $N—CH_2—CH(OH)—CH(CO_2H)—NH—CH_2—CH_2—C(NH_2)CO_2H$	*Hordeum vulgaris*	*184b*
83	4-Hydroxyornithine $H_2N—CH_2—CH(OH)—CH_2—CH(NH_2)CO_2H$	*Vicia sativa* *Vicia onobrychoides* *Vicia unijuga*	*10* *11* *11*
84	2-Amino-4-hydroxypentanoic acid (4-hydroxynorvaline) $H_3C—CH(OH)—CH_2—CH(NH_2)CO_2H$	*Lathyrus odoratus*	*52*
85	2-Amino-5-hydroxypentanoic acid (5-hydroxynorvaline) $HO—CH_2—CH_2—CH_2—CH(NH_2)CO_2H$	*Glycine max* *Canavalia ensiformis*	*189* *11* *11*

216

#	Compound	Source	Ref.
86	2-Amino-4,5-dihydroxypentanoic acid HO—CH₂—CH(OH)—CH₂—CH(NH₂)CO₂H	*Lunaria annua*	116
87	2-Amino-4-hydroxypimelic acid HO₂C—(CH₂)₂—CH(OH)—CH₂—CH(NH₂)CO₂H	*Asplenium septentrionale* 2(*S*): *Reseda luteola*	210
88	3-Hydroxypipecolic acid 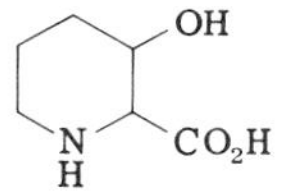	*Peganum harmala*	72
89	*trans*-4-Hydroxypipecolic acid 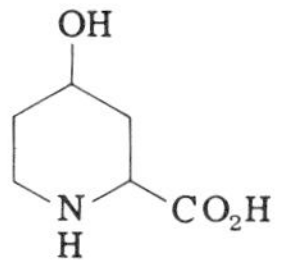	*Acacia pentadenia* Mimosaceae *cis: Strophantus scandens*	22 203 164b
90	*trans*-5-Hydroxypipecolic acid 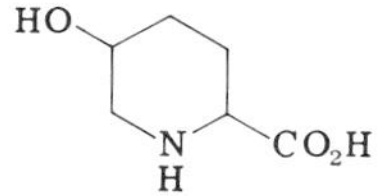	*Rhapis flabelliformis* *Baikiaea plurijuga* *Acacia* *cis: Gleditsia triacanthos*	202 79 42 188
91	5-Hydroxy-6-methylpipecolic acid 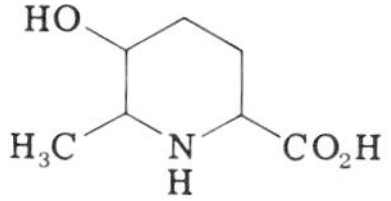	(2*S*, 5*S*, 6*S*): *Fagus sylvatica* (2*S*, 5*R*, 6*S*): *Fagus sylvatica*	110 110

$$HO—CH_2—CH(OH)—CH_2—CH(NH_2)CO_2H$$

$$HO_2C—(CH_2)_2—CH(OH)—CH_2—CH(NH_2)CO_2H$$

(structures for 88, 89, 90, 91: pipecolic acid ring structures with OH/CH₃ substituents and CO₂H groups)

(continued)

APPENDIX (Continued)

No.	Compound	Source	Reference
92	4,5-Dihydroxypipecolic acid 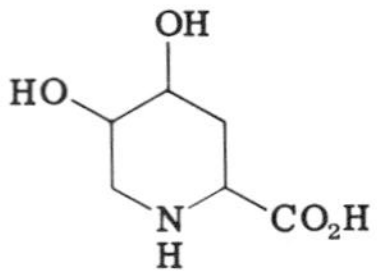 	Calliandra haematocephala (4R, 5R): Julbernardia paniculata (4S, 5S): Derris elliptica (4R, 5S): Derris elliptica	126 169b 169b 169b
93	trans-3-Hydroxyproline 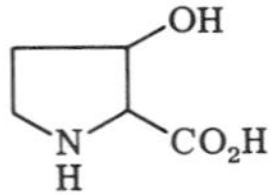 	Delonix regia	177
94	trans-4-Hydroxyproline* 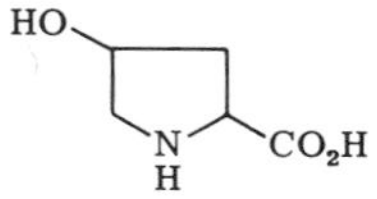 	cis: Santalum album	154
95	trans-4-Hydroxymethylproline 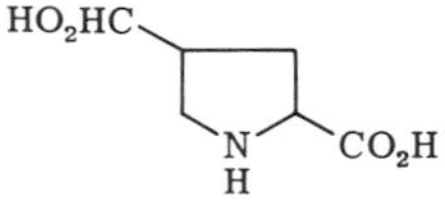 	Eriobotrya japonica cis: Rosaceae	73b 50

No.	Compound	Source	Ref.
96	3-(2,6-Dihydroxypyrimidin-5-yl)alanine 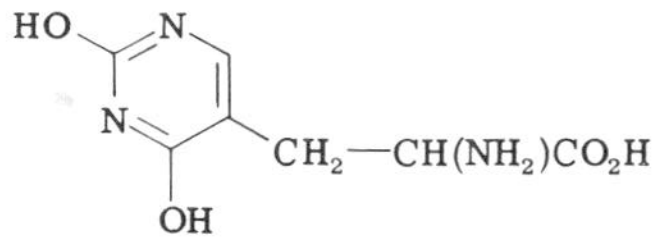 	*Pisum sativum*	*17*
97	O-Acetylserine $H_3C-C(=O)-O-CH_2-CH(NH_2)COOH$	*Nicotinia*	*171b*
98	O-Phosphoserine $H_2O_3P-O-CH_2-CH(NH_2)CO_2H$	*Phaseolus vulgaris*	*26b*
99	5-Hydroxytryptophan 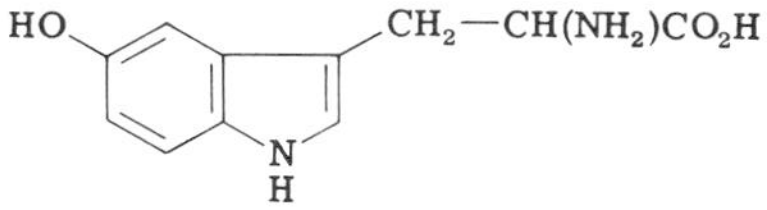 	*Griffonia simplicifolia*	*7*
100	4-Hydroxyvaline 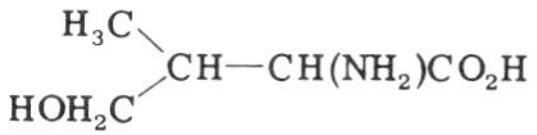 	*Kalanchoe diagremontiana*	*151*

*See **162**

(continued)

APPENDIX (*Continued*)

No.	Compound	Source	Reference
Group VI: Ureido and guanidino acids and similar derivatives			
101	Albizziine (2-amino-3-ureidopropionic acid) $H_2N—C(=O)—NH—CH_2—CH(NH_2)CO_2H$	*Acacia* *Albizzia lophanta* *Albizzia julibrissin* Mimosaceae	*42* *42* *70* *144*
102	Argininosuccinic acid (N-{[(4-amino-4-carboxybutyl)—amino]-iminomethyl}aspartic acid) 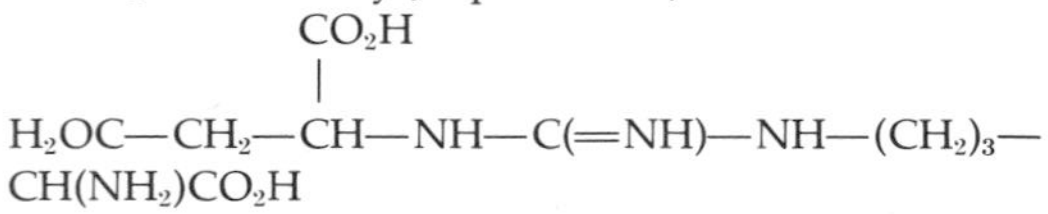 CO_2H $\|$ $H_2OC—CH_2—CH—NH—C(=NH)—NH—(CH_2)_3—CH(NH_2)CO_2H$	May be widely distributed	
103	Canavanine [2-Amino-4-(guanidinooxy)butyric acid] $H_2N—C(=NH)—NH—O—CH_2—CH_2—CH(NH_2)$ CO_2H	Leguminosae	*162* *106*
104	Canavaninosuccinic acid (N-{[(3-amino-3-carboxypropoxy)amino]-iminomethyl}aspartic acid) 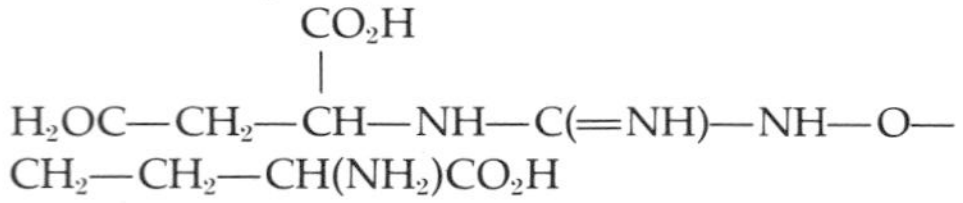 CO_2H $\|$ $H_2OC—CH_2—CH—NH—C(=NH)—NH—O—$ $CH_2—CH_2—CH(NH_2)CO_2H$	Leguminosae	*162* *214*
105	Citrulline (2-amino-5-ureidopentanoic acid) $H_2N—C(=O)—NH—(CH_2)_3—CH(NH_2)CO_2H$	*Citrullus vulgaris* May be widely distributed	*213*
106	Deaminocanavanine (hexahydro-3-imino-1,2,4-oxadia-zepine-5-carboxylic acid) 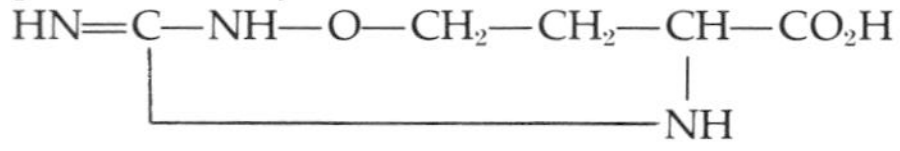 $HN=C—NH—O—CH_2—CH_2—CH—CO_2H$ $—NH$	Leguminosae*	

107	Gigartinine [2-amino-5-(guanidoureido)pentanoic acid or N^5-(amidinocarbamoyl)-ornithine] $H_2N—C(=NH)—NH—C(=O)—NH—(CH_2)_3—CH(NH_2)CO_2H$	*Gymnogongrus flabelliformis*	*96*
108	Homoarginine (2-amino-6-guanidinohexanoic acid) $H_2N—C(=NH)—NH—(CH_2)_4—CH(NH_2)CO_2H$	*Lathyrus cicera* *Lathyrus sativus*	*5* *155*
109	Homocitrulline (2-amino-6-ureidohexanoic acid) $H_2N—C(=O)—NH—(CH_2)_4—CH(NH_2)CO_2H$	*Vicia faba*	*94*
110	O-Ureidohomoserine [2-amino-4-(ureidooxy)butyric acid] $H_2N—C(=O)—NH—O—CH_2—CH_2—CH(NH_2)—CO_2H$	*Leguminosae**	
111	Indospicine (2-amino-6-amidinohexanoic acid) $H_2N—C(=NH)—(CH_2)_4—CH(NH_2)CO_2H$	*Indigofera spicata*	*84*

Group VII: Aromatic constituents

112	6-Hydroxykynurenine 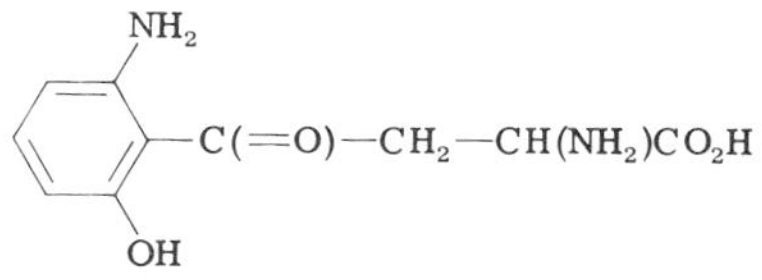	*Nicotinia*	*123*

113	3-(4-Aminophenyl)alanine 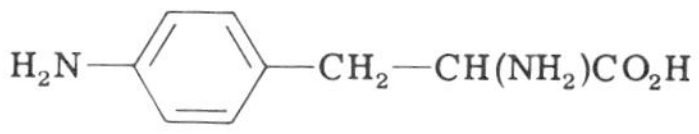	*Vigna vexillata*	*29*

---

* Canavanine metabolic product; may be present in canavanine-containing legumes.

(continued)

APPENDIX (*Continued*)

No.	Compound	Source	Reference
114	3-(3-Aminomethylphenyl)alanine 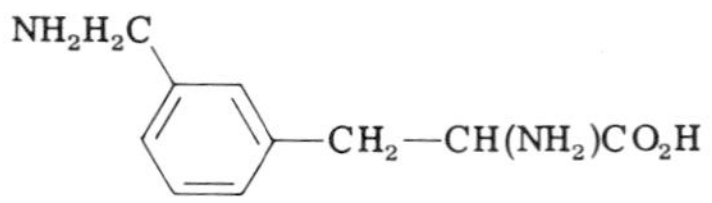 NH_2H_2C —⟨ring⟩— CH_2—$CH(NH_2)CO_2H$	*Combretum zeyheri*	*140*
115	3-(3-Carboxyphenyl)alanine 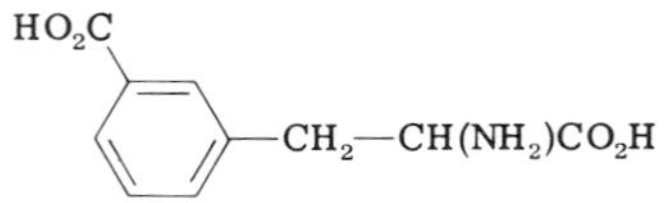 HO_2C —⟨ring⟩— CH_2—$CH(NH_2)CO_2H$	*Reseda luteola* *Curcubitaceae* *Iris tingitana*	*105* *34* *190*
116	3-(3-Hydroxyphenyl)alanine (*m*-tyrosine) 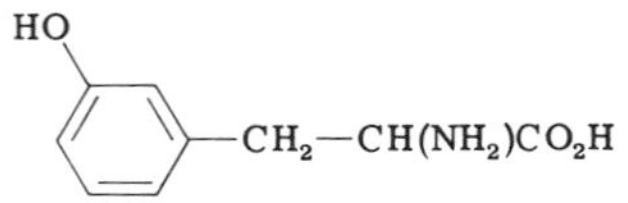 HO —⟨ring⟩— CH_2—$CH(NH_2)CO_2H$	*Euphorbia myrinitis*	*137*
117	3-(3-Carboxy-4-hydroxyphenyl)alanine (3-carboxytyrosine) 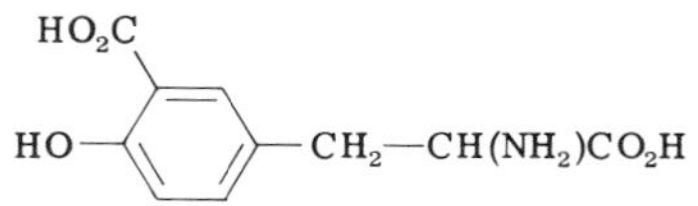 HO_2C / HO —⟨ring⟩— CH_2—$CH(NH_2)CO_2H$	*Reseda luteola*	*105*
118	3-(3-Hydroxymethylphenyl)alanine 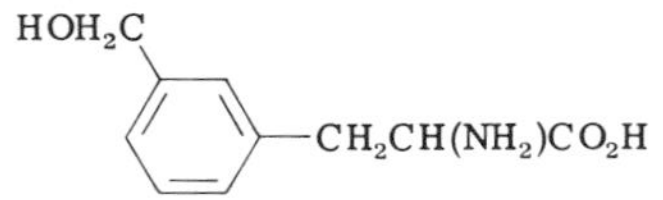 HOH_2C —⟨ring⟩— $CH_2CH(NH_2)CO_2H$	*Caesalpinia tinctoria*	*215*

119	3-(3-Hydroxymethyl-4-hydroxyphenyl)alanine 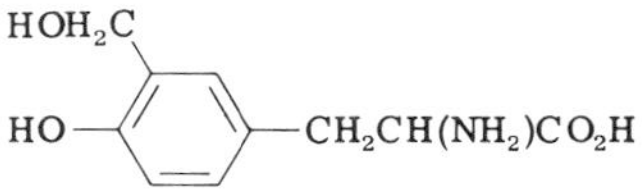	*Caesalpinia tinctoria*	215

$$\text{HOH}_2\text{C}$$
$$\text{HO} \quad \text{CH}_2\text{CH(NH}_2)\text{CO}_2\text{H}$$

120 3-(2,4-Dihydroxy-6-methylphenyl)alanine (β-orcylalanine)   *Agrostemma githago*   165

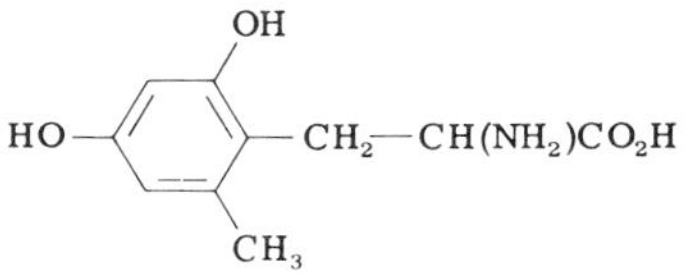

$$\text{OH}$$
$$\text{CH}_2\text{—CH(NH}_2)\text{CO}_2\text{H}$$
$$\text{HO} \quad \text{CH}_3$$

121a 3-(3,4-Dihydroxyphenyl)alanine (dopa)   *Vicia faba*   192

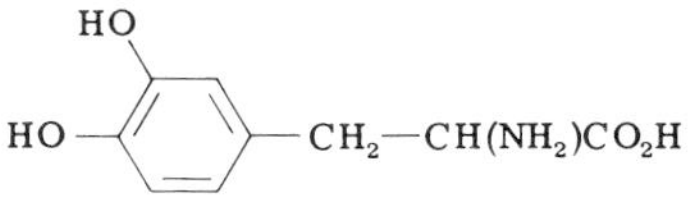

$$\text{CH}_2\text{—CH(NH}_2)\text{CO}_2\text{H}$$
$$\text{HO}$$
$$\text{HO}$$

121b N^5-(2-hydroxybenzyl)-4-hydroxyglutamine   (2S, 4S): *Fagopyrum esculentum*   108b

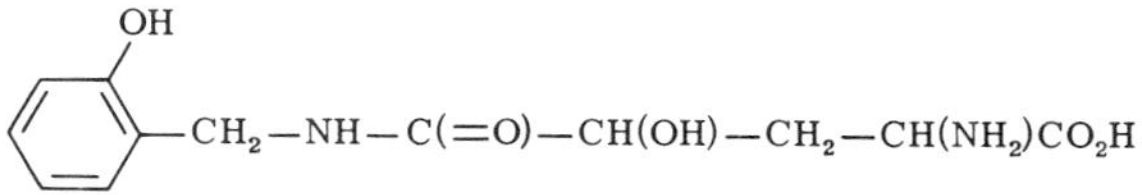

$$\text{OH}$$
$$\text{CH}_2\text{—NH—C(=O)—CH(OH)—CH}_2\text{—CH(NH}_2)\text{CO}_2\text{H}$$

122 2-(Phenyl)glycine   *Fagus sylvatica*   32

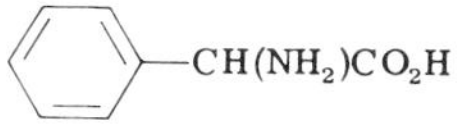

$$\text{CH(NH}_2)\text{CO}_2\text{H}$$

(continued)

APPENDIX (*Continued*)

No.	Compound	Source	Reference
123	2-(3-Carboxyphenyl)glycine 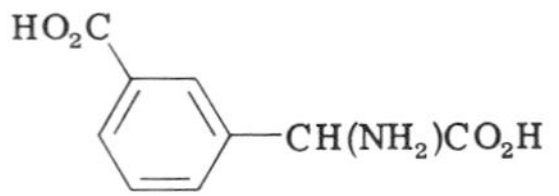 HO_2C — (benzene ring) — $CH(NH_2)CO_2H$	*Reseda luteola* *Iris tingitana*	*105* *190*
124	2-(3-Hydroxyphenyl)glycine 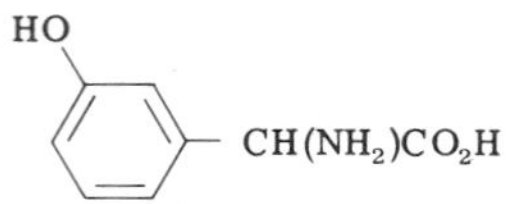 HO — (benzene ring) — $CH(NH_2)CO_2H$	*Euphorbia helioscopia*	*138*
125	2-(3-Carboxy-4-hydroxyphenyl)glycine 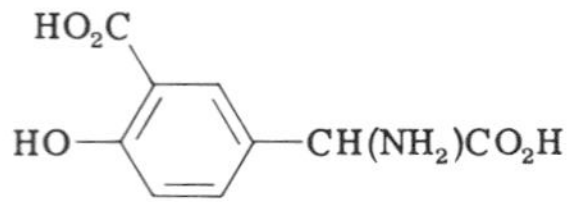 HO_2C, HO — (benzene ring) — $CH(NH_2)CO_2H$	*Reseda luteola*	*105*
126a	2-(3,5-Dihydroxyphenyl)glycine 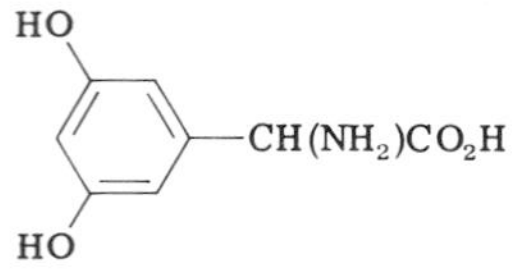 HO, HO — (benzene ring) — $CH(NH_2)CO_2H$	*Euphorbia helioscopa*	*138*
126b	N^5-(Benzoyl)-ornithine 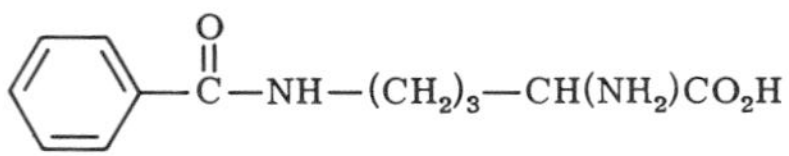 (phenyl)—$\overset{O}{\overset{\|}{C}}$—$NH$—$(CH_2)_3$—$CH(NH_2)CO_2H$	*Vicia pseudo-orobus*	*82b*

126c	N^5-(Benzoyl)-4-hydroxyornithine	Vicia pseudo-orobus	82b
	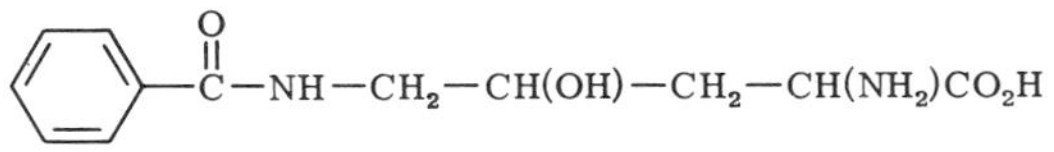 $C_6H_5-C(=O)-NH-CH_2-CH(OH)-CH_2-CH(NH_2)CO_2H$		
127	3,5,3'-Triiodothyronine	Phaseolus vulgaris	48
		Several monocots	
128	3,5-Dibromotyrosine	Laminaria flexicaulis	161
129	Monoiodotyrosine (3'-iodotyrosine)	Nereocystis leutkeana	191
		Ulva lactuca	191
		Laminaria digitata	167
130	3,5-Diiodotyrosine	Laminaria flexicaulis	161

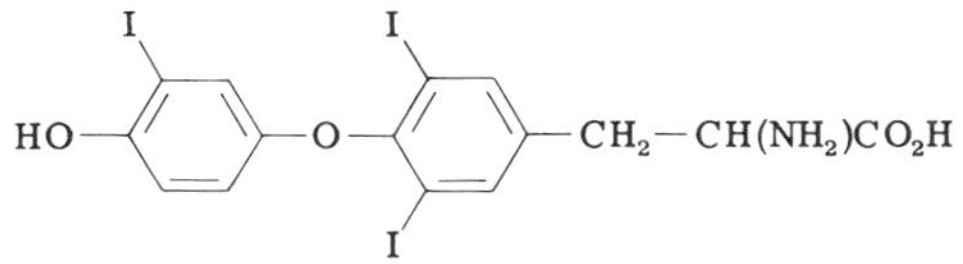

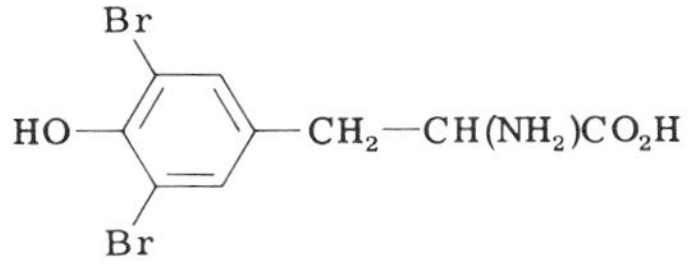

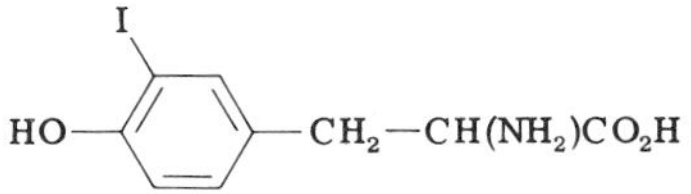

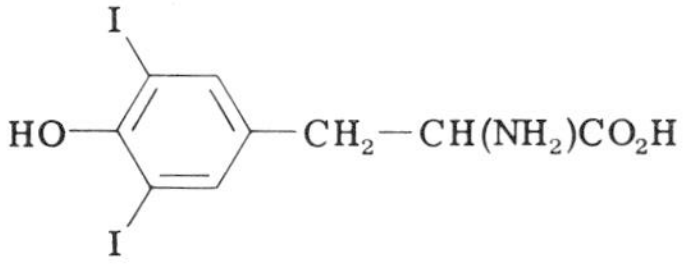

(continued)

APPENDIX (*Continued*)

No.	Compound	Source	Reference
Group VIII: Heterocyclic compounds			
131	4-Aminopipecolic acid 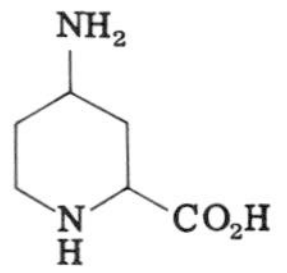	*Strophanthus scandens*	*164*
132	Ascorbalamic acid 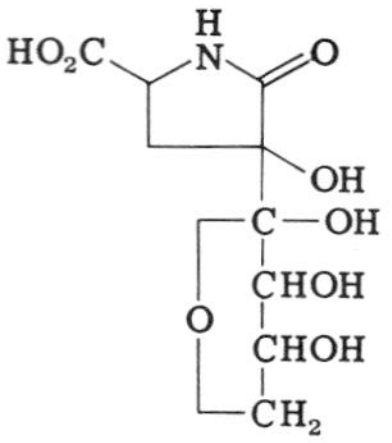	May be widely distributed	23
133	Azetidine-2-carboxylic acid 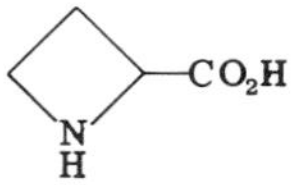	*Polygonatum officinale* *Delonix regia* *Beta vulgaris* *Convallaria majalis*	*204* *177* *53* *46*
134	*N*-(3-amino-3-carboxypropyl)azetidine-2-carboxylic acid 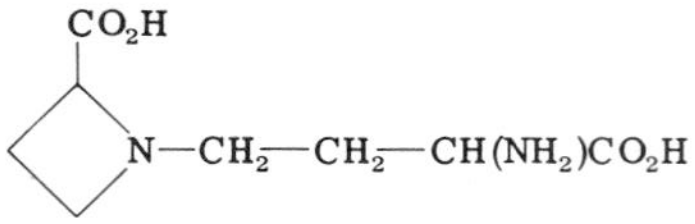	(2*S*, 3*S*): *Fagus sylvatica*	*109*

#	Name / Structure	Source	Refs.
135*	N-[N-(3-amino-3-carboxypropyl)-3-amino-3-carboxy-propyl]azetidine-2-carboxylic acid (nicotianimine) 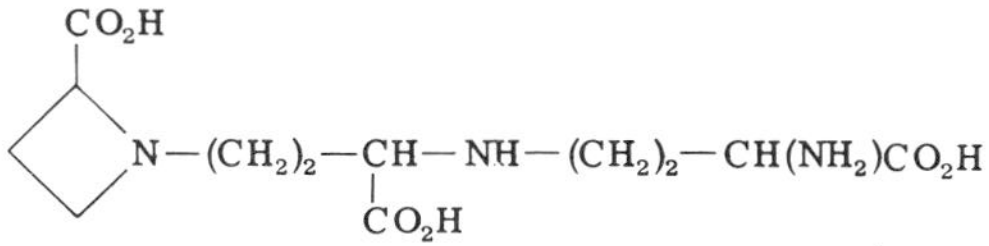 $N-(CH_2)_2-CH-NH-(CH_2)_2-CH(NH_2)CO_2H$ with CO_2H substituent on the azetidine ring bearing CO_2H	*Fagus sylvatica* *Nicotiana tabacum*	*109* *143b*
136	Baikiain (4,5-dehydropipecolic acid) 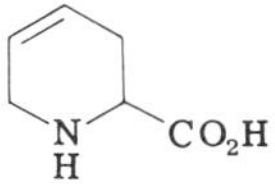 (ring structure with CO_2H, N–H)	*Baikiaea plurijuga* *Corallina officinalis* *Caesalpinia tinctoria*	*104* *124* *215*
137	3-Carboxy-1,2,3,4-tetrahydro-3-carboline 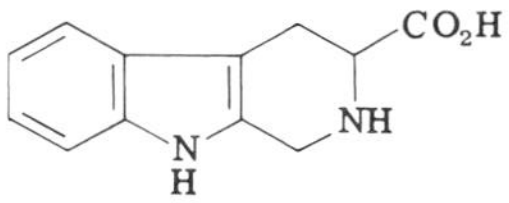 (fused ring structure with CO_2H, NH, N–H)	*Aleurites fordii*	*145*
138	Cucurbitine (3-amino-3-carboxypyrrolidine) 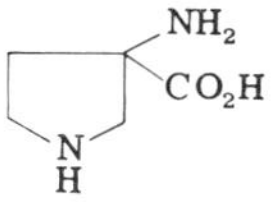 (ring structure with NH_2, CO_2H, N–H)	*Cucurbita moschata*	*43*
139	2-(Cyclopent-2'-enyl)glycine 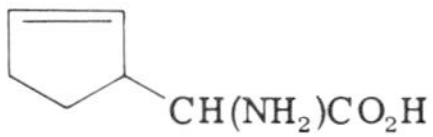 (cyclopentene with $CH(NH_2)CO_2H$)	(1'R, 2S): *Hydnocarpus anthelminthica* (1'S, 2S): *Caloncoba echinata*	*24*

* See compound **82**.

(*continued*)

APPENDIX (*Continued*)

No.	Compound	Source	Reference
140	Domoic acid [2-carboxy-3-carboxymethyl-4-(1-methyl-2-carboxy-1,3-hexadienyl)pyrrolidine] 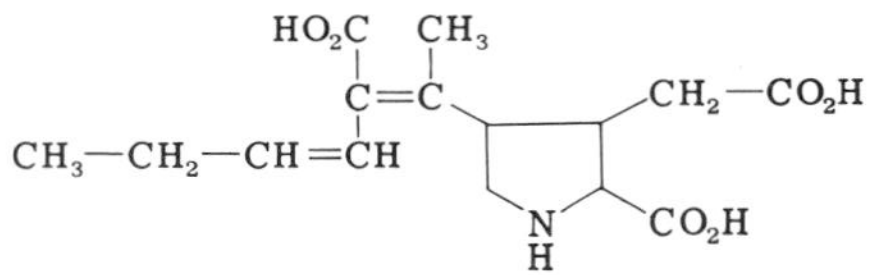	*Chondria armata*	26
141	Enduracididine [3-(2-Amino-2-imidazolin-4-yl)alanine] 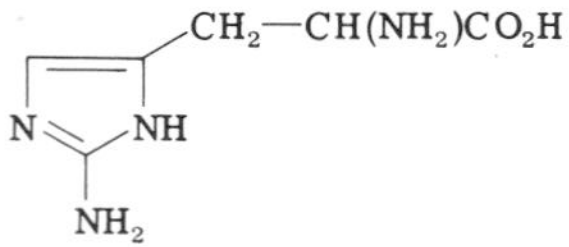	*Lonchocarpus sericeus*	44
142	3-(2-Furoyl)alanine 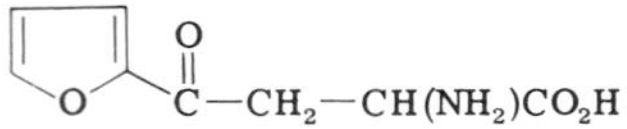	*Fagus sylvatica* *Fagopyrum esculentum*	110 89
143	Guvacine 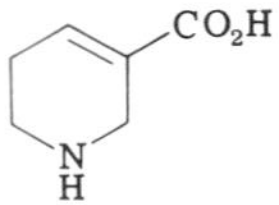	*Areca cathechu*	101
144	τ-Methylhistidine 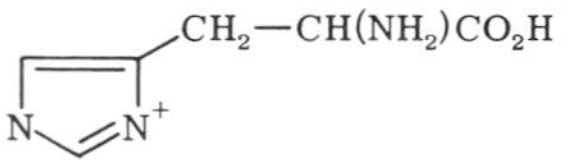	*Phyllospora comosa*	124

| 145 | π,τ-Dimethylhistidine 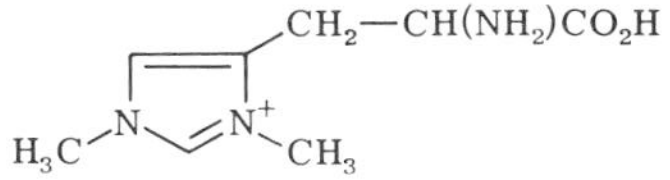 | *Gracilaria secundata* | 124 |

$$CH_2-CH(NH_2)CO_2H$$

(structure: imidazolium ring with H_3C-N and N^+-CH_3)

| 146 | N-(Indole-3-acetyl)aspartic acid 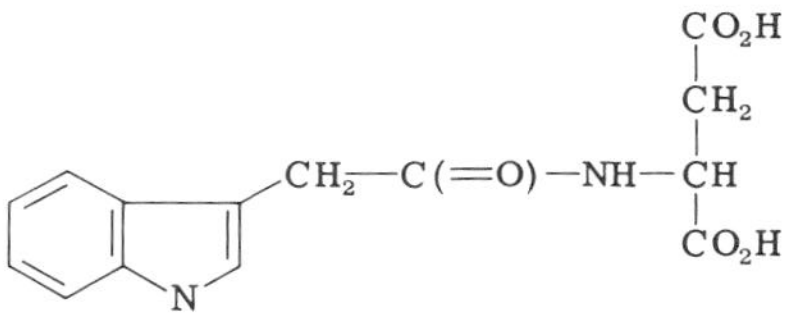 | *Magnolia* | 212 |
| | | *Pisum* | 2 |

$$CH_2-C(=O)-NH-CH$$

with CO_2H, CH_2, CO_2H side chain; indole ring (N)

| 147 | Isowillardiine [2-amino-3,6-dihydro-2,6-dioxo-1(2H)pyrimidinepropionic acid] 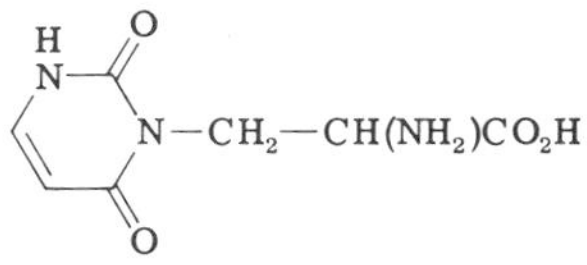 | *Pisum* | 113 |

(structure: pyrimidinedione ring with $N-CH_2-CH(NH_2)CO_2H$)

| 148 | 3-(Isoxazolin-5-one-2-yl)alanine 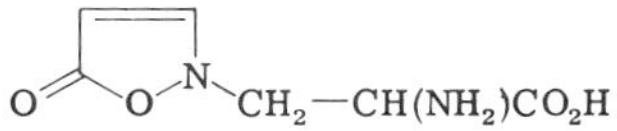 | *Pisum sativum* | 114 |
| | | *Lathyrus odoratus* | |

$$N-CH_2-CH(NH_2)CO_2H$$

(isoxazolinone ring)

| 149 | 3-(2-β-D-Glucopyranosylisoxazolin-5-one-4-yl)alanine 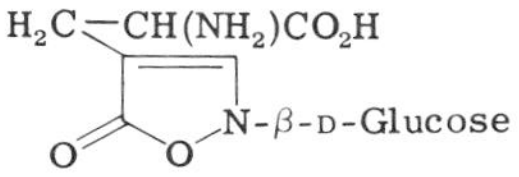 | *Pisum sativum* | 114 |
| | | *Lathyrus odoratus* | |

$$H_2C-CH(NH_2)CO_2H$$

with $N-\beta$-D-Glucose (isoxazolinone ring)

(continued)

APPENDIX (*Continued*)

No.	Compound	Source	Reference
150	2-Amino-4-(isoxazolin-5-one-2-yl)butyric acid 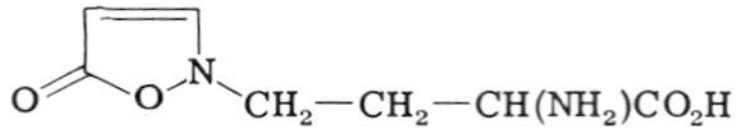 $O\!\!=\!\!\overset{\displaystyle N}{\underset{O}{\bigcirc}}\!\!-CH_2-CH_2-CH(NH_2)CO_2H$	*Lathyrus odoratus*	*114*
151	α-Kainic acid 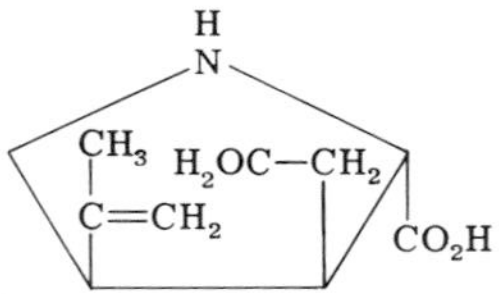 (structure)	*Digenea simplex*	*194* *185*
152	Lathyrine 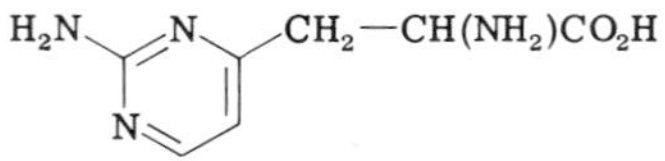 $H_2N\!\!-\!\!\overset{N}{\underset{N}{\bigcirc}}\!\!-CH_2-CH(NH_2)CO_2H$	*Lathyrus tingitanus*	*4*
153	Tetrahydrolathyrine 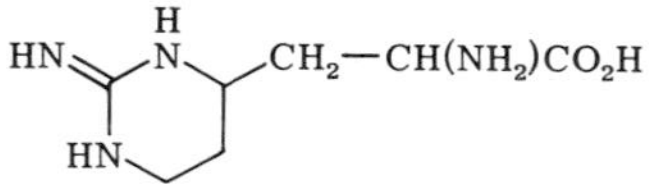 (structure) $-CH_2-CH(NH_2)CO_2H$	*Lonchocarpus costaricensis*	*45*
154	Nicotianine [*N*-(3-amino-3-carboxypropyl)-3-carboxy-pyridinium betaine] 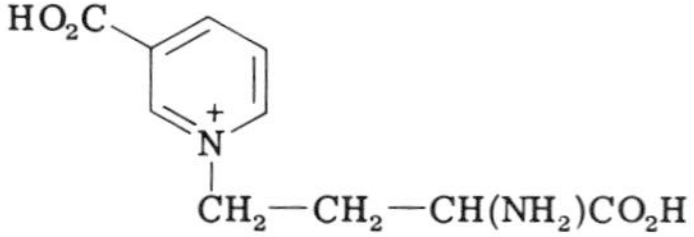 $HO_2C\!\!-\!\!\overset{+}{\underset{}{\bigcirc}}\!\!-CH_2-CH_2-CH(NH_2)CO_2H$	*Nicotiana tabacum*	*144*

155	Penmacric acid 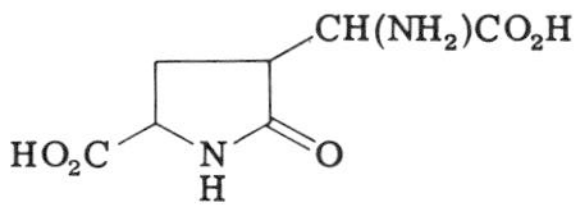	*Pentaclethra macrophylla*	*130*
156	Pipecolic acid (piperidine-2-carboxylic acid) 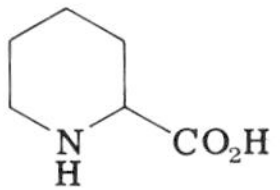	Probably widely distributed	*80* *77*
157	*trans*-4-Carboxyproline 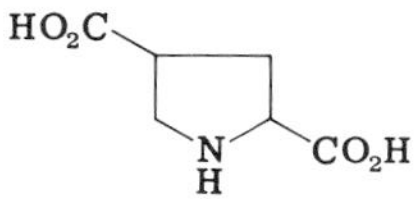	*Afzelia bella* *Chondria coerulescens*	*216* *87* *191*
158a	5-Carboxyproline 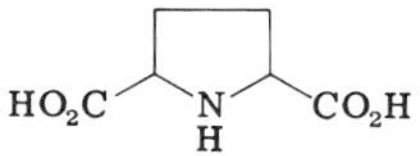	*Schizymenia dubyi*	*90*
158b	2,4-Methanoproline (2-carboxy-2,4-methanopyrrolidine) 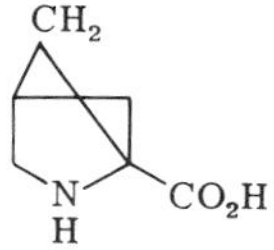	*Ateleia herbert smithii*	*12b*
159	*exo-cis*-3,4-Methanoproline 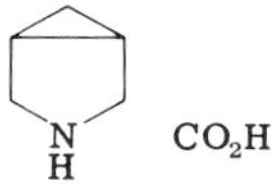	*Aesculus parviflora*	*60*

(continued)

231

APPENDIX (Continued)

No.	Compound	Source	Reference
160	trans-4-Methylproline 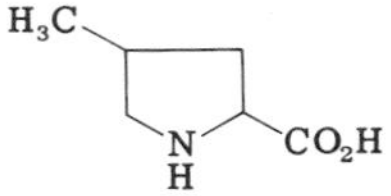	Malus	88
161	1-Methylproline (hygric acid) 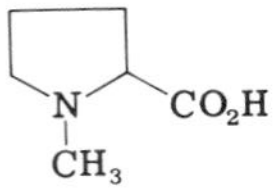	Malus	88
162	1-Methyl-4-hydroxyproline 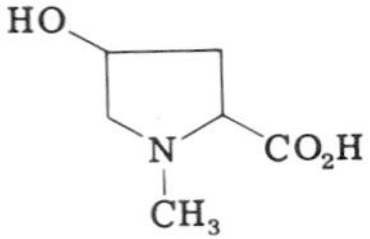	Croton megalobotrys Afrormosia elata	73 134
163	4-Methyleneproline 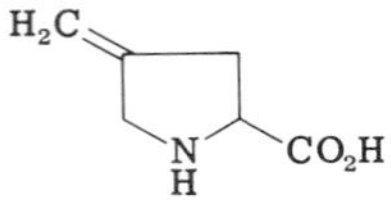	Eriobotrya japonica	76
164	3-Pyrazol-1-ylalanine 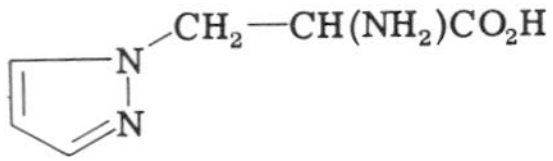	Citrullus vulgaris	143
165	Quisqualic acid [3-(3,5-dioxo-1,2,4-oxazolidin-2-yl)ala-nine] 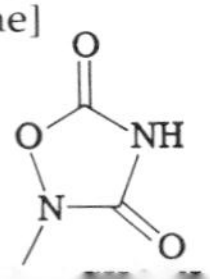	Quisqualis indica	139

166	Stizolobic acid (2-amino-6-carboxy-2-oxo-2H-pyran-4-propionic acid) 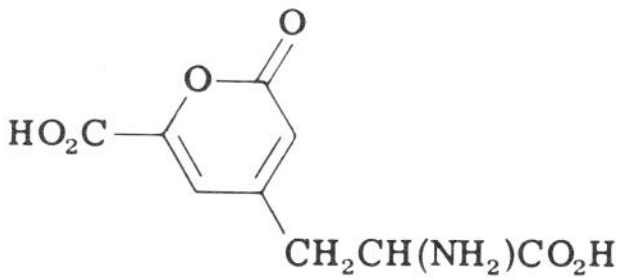	Stizolobium hassjoo	83
167	Stizolobinic acid (2-amino-6-carboxy-2-oxo-2H-pyran-3-propionic acid) 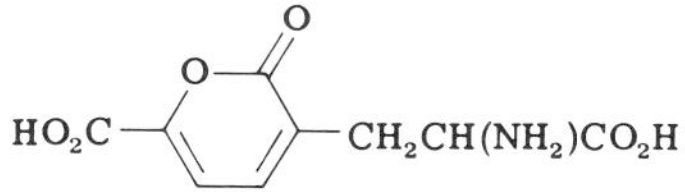	Stizolobium hassjoo	169
168	Trigonelline (coffearin) 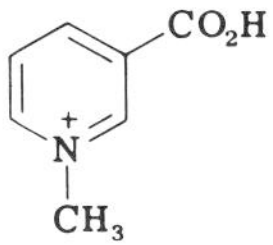	Trigonella	100
169	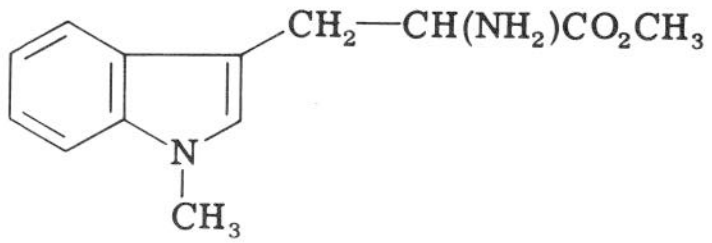 N¹-Methyltryptophan (methyl ester)	Aotus subglauca	102
170	Willardiine (2-amino-3,4-dihydro-2,4-dioxo-1(2H)-pyrimidine propionic acid) 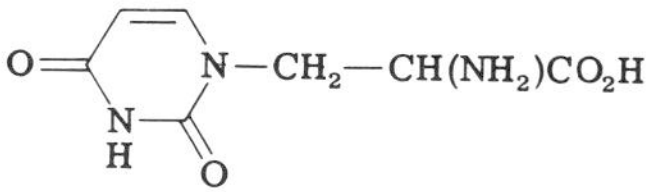	Mimosa	62
		Acacia willardiana	62
		Pisum	113
		Acacia	42

(continued)

APPENDIX (*Continued*)

No.	Compound	Source	Reference
Group IX: N-Substituted derivatives			
171	*N*-Carboethoxyacetyl-D-4-chlorotryptophan 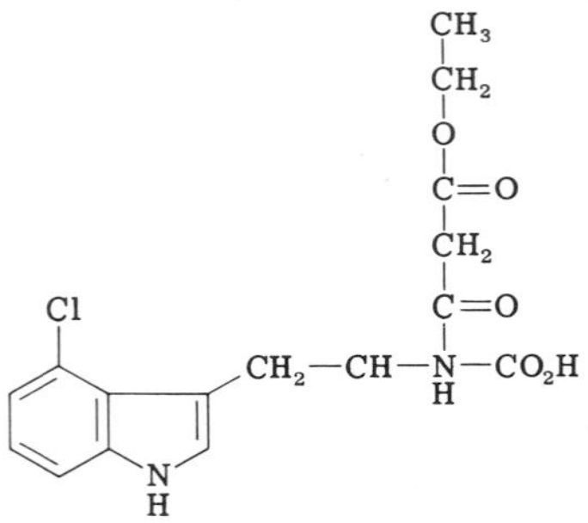	*Pisum sativum*	127
172	*N*-Carbomethoxyacetyl-D-4-chlorotryptophan 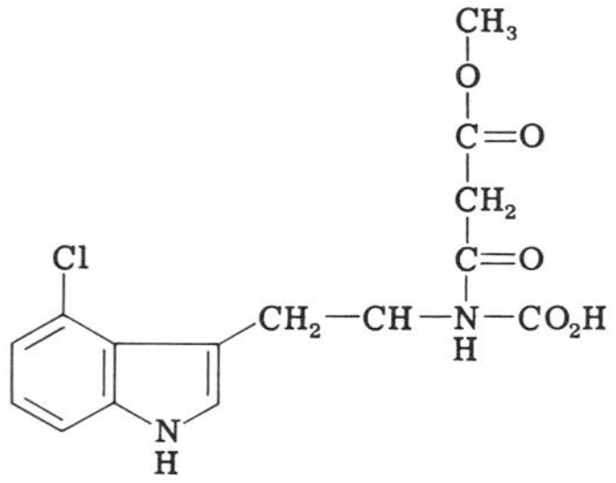	*Pisum sativum*	127

| 173 | Abrine (*N*-methyltryptophan) 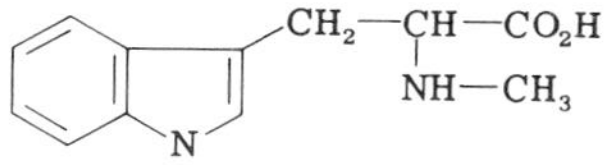 | *Abrus precatorius* | 87 |

$$CH_2-CH-CO_2H$$
$$NH-CH_3$$

(indole structure)

174	*N*-Methylalanine $H_3C-CH(NHCH_3)CO_2H$	*Dichapetalum cymosum*	39
175a	N^2-Oxalyl-2,4-diaminobutyric acid $H_2N-CH_2-CH_2-CH(HNCOCO_2H)CO_2H$	*Lathyrus latifolius*	8
175b	N^2-Oxalyl-2,3-diaminopropionic acid $H_2N-CH_2-CH(HN-COCO_2H)CO_2H$	*Lathyrus sativus*	8
176	*N*-(2-Hydroxyethyl)glycine (petalonine) CH_2-CH_2-OH $H_2C-NH-CO_2H$	*Petalonia fasica*	182
177	*N*-Carbomoyl-2-(4-hydroxyphenyl)glycine 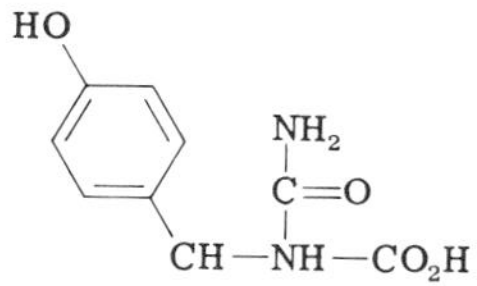	*Vicia faba*	36

$$NH_2$$
$$C=O$$
$$CH-NH-CO_2H$$
HO

178	*N*-Acetylhomoserine $HO-CH_2-CH_2-CH(NHCOCH_3)CO_2H$	May be an isolation artifact rather than a natural product	
179	*N*-Formylmethionine $H_3C-S-CH_2-CH_2-CH(NH-CHO)CO_2H$	Chloroplasts	166
180	Octopine [*N*-(D-1-carboxyethyl)-arginine] $H_2N-C(=NH_2)-NH-(CH_2)_3-CH-N-CO_2H$ $H_3C-CH-CO_2H$	Various leaf and stem tissues	

(continued)

APPENDIX (*Continued*)

No.	Compound	Source	Reference
181	Sarcosine (*N*-methylglycine) $H_2C(NHCH_3)CO_2H$	*Cladonia silvatica*	*121*
182	*N*-Methylserine $HO-CH_2-CH(NHCH_3)CO_2H$	*Dichapetalum cymosum*	*40*
183	*N,N* -Dimethyltryptophan 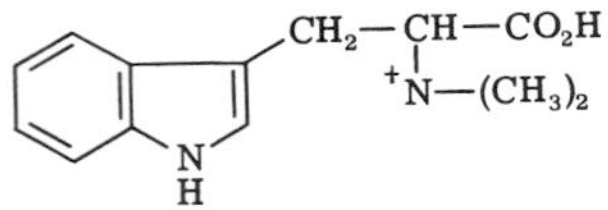	*Abrus precatorius*	*124b*
184	*N*-Methyltyrosine 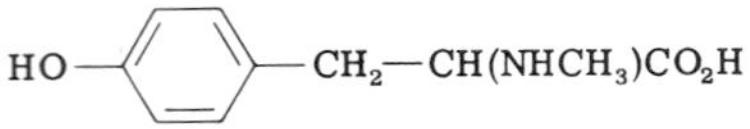	*Andira* (bark) *Combretum zeyheri*	*140b*

Group X: Sulfur-containing compounds

No.	Compound	Source	Reference
185	Alliin (*S*-allylcysteine sulfoxide) $H_2C=CH-CH_2-S-CH_2-CH(NH_2)CO_2H$	*Allium ursinum* *Allium sativum*	*175* *175*
186	Chondrine (1,4-thiazane-3-carboxylic acid 1-oxide) 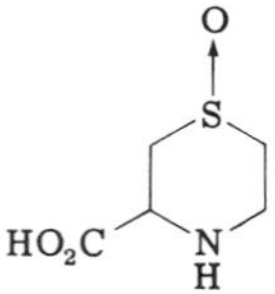	*Chondria crassicaulis* *Undaria pinnatifida*	*112*

187	Cycloalliin (3-methyl-1,4-thiazane-5-carboxylic acid 1-oxide) 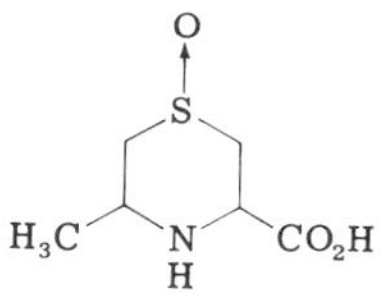	Allium spp.	*205*
188	Cystathionine (2-amino-4-[(2-amino-2-carboxyethyl)thio]butyric acid) CH_2—S—CH_2 | | CH_2 $CH(NH_2)CO_2H$ | $CH(NH_2)CO_2H$	May be widespread	*30*
189	S-Allylcysteine $H_2C{=}CH{-}CH_2{-}S{-}CH_2{-}CH(NH_2)CO_2H$	Allium cepa Allium sativum	*156* *156*
190	S-Allylmercaptocysteine $H_2C{=}CH{-}CH_2{-}S{-}S{-}CH_2{-}CH(NH_2)CO_2H$	Allium sativum	*176*
191a	S-(2-Carboxyethyl)cysteine $HO_2C{-}(CH_2)_2{-}S{-}CH_2{-}CH(NH_2)CO_2H$	Albizzia julibrissin Acacia	*70* *42*

191b	S-(1,2-Dicarboxyethyl)cysteine
	$H_2OC{-}CH_2{-}CH{-}S{-}CH_2{-}CH(NH_2)CO_2H$
	              |
	            CO_2H
		Asparagus officinalis	*102c*

192	S-(2-Carboxyisopropyl)cysteine
	$CH_3{-}CH{-}CH_2{-}CO_2H$
	     |
	     S
	     |
	  $CH_2{-}CH(NH_2)CO_2H$
		Acacia
		Albizzia lebek
			42
			64
			19

193	S-(Carboxymethyl)cysteine $HO_2C{-}CH_2{-}S{-}CH_2{-}CH(NH_2)CO_2H$	Raphanus sativus	*20*

(continued)

237

APPENDIX (*Continued*)

No.	Compound	Source	Reference
194	S-(2-Carboxypropyl)cysteine 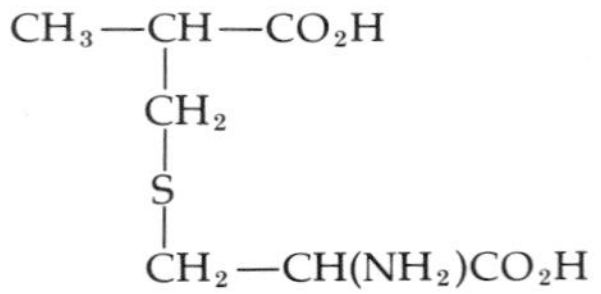 CH_3—CH—CO_2H 　　| 　　CH_2 　　| 　　S 　　| 　　CH_2—CH(NH_2)CO_2H	*Allium cepa*	*112* *207*
195	Homocysteine HS—CH_2—CH_2—CH(NH_2)CO_2H	*Spinacia*	*180*
196	S-Methylcysteine CH_3—S—CH_2—CH(NH_2)CO_2H	*Phaseolus vulgaris*	*107*
197	S-n-Propylcysteine CH_3—$(CH_2)_2$—S—CH_2—CH(NH_2)CO_2H	*Allium sativum*	*209*
198	S-(Prop-1-enyl)cysteine CH_3—CH=CH—S—CH_2—CH(NH_2)CO_2H	*Allium sativum*	*128*
199	3,3'-(2-Methylethylene-1,2-dithio)dialanine 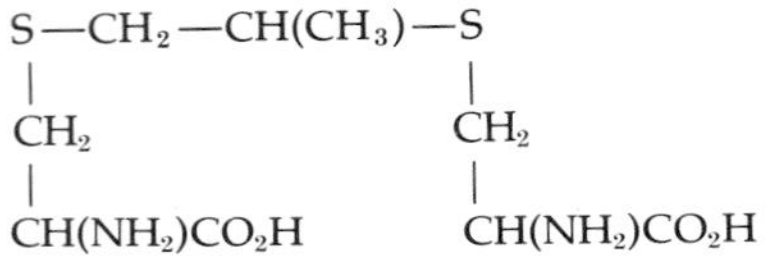 S—CH_2—CH(CH_3)—S |　　　　　　　| CH_2　　　　　CH_2 |　　　　　　　| CH(NH_2)CO_2H　CH(NH_2)CO_2H	*Allium shoenoprasum*	*129*
200	Djenkolic acid [3,3'-methylenedithiobis(2-amino-propionic acid)] 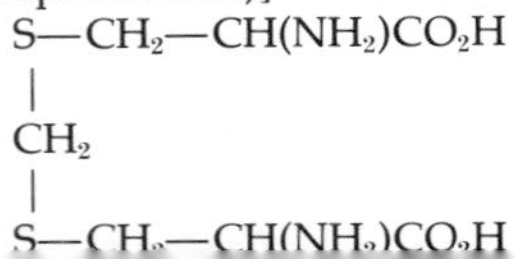 S—CH_2—CH(NH_2)CO_2H | CH_2 | S—CH_2—CH(NH_2)CO_2H	*Acacia* *Pithecolobium lobatum* *Albizzia lophantha*	*42* *196* *68*

238

201	N^6-Acetyldjenkolic acid 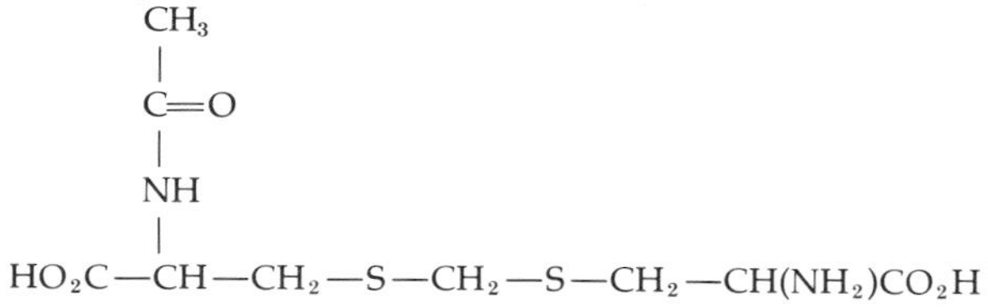 CH_3 $\mid$ $C{=}O$ $\mid$ NH $\mid$ $HO_2C{-}CH{-}CH_2{-}S{-}CH_2{-}S{-}CH_2{-}CH(NH_2)CO_2H$	*Acacia* *Acacia farnesiana*	42 69
202	Homomethionine $CH_3{-}S{-}(CH_2)_3{-}CH(NH_2)CO_2H$	*Brassica*	179
203	S-Methylmethionine $(CH_3)_2{-}S^+{-}CH_2{-}CH_2{-}CH(NH_2)CO_2^-$	*Asparagus*	21
204	S-(2-Carboxyethyl)cysteine sulfoxide*	*Acacia*	42
205	S-Methylcysteine sulfoxide 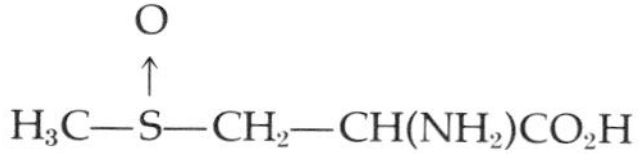 $\quad\quad\quad\quad O$ $\quad\quad\quad\quad\uparrow$ $H_3C{-}S{-}CH_2{-}CH(NH_2)CO_2H$	*Allium cepa* *Brassica rapa* *Phaseolus*	206 135 19
206	S-*n*-Propylcysteine sulfoxide (dihydroalliin) 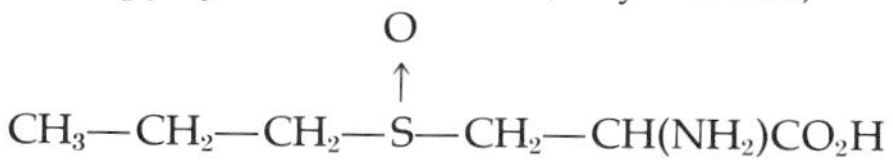 $\quad\quad\quad\quad\quad\quad\quad O$ $\quad\quad\quad\quad\quad\quad\quad\uparrow$ $CH_3{-}CH_2{-}CH_2{-}S{-}CH_2{-}CH(NH_2)CO_2H$	*Allium cepa*	206
207	S-(Prop-1-enyl)cysteine sulfoxide 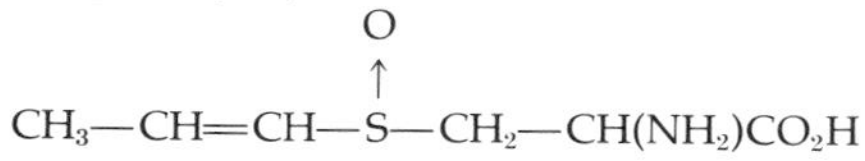 $\quad\quad\quad\quad\quad\quad\quad O$ $\quad\quad\quad\quad\quad\quad\quad\uparrow$ $CH_3{-}CH{=}CH{-}S{-}CH_2{-}CH(NH_2)CO_2H$	*Allium sativum*	172
208	Djenkolic acid sulfoxide	*Acacia georginae* *Acacia*	168 42
209	N-Acetyldjenkolic acid sulfoxide*	*Acacia* spp.	42

* May be artifacts arising from oxidation of the parent compound (see ref. 160).

(continued)

APPENDIX (*Continued*)

No.	Compound	Source	Reference
210	Taurine (2-aminoethanesulfonic acid) $H_2N-CH_2-CH_2-SO_3H$	Red algae	111
211	N-(1-Carboxyethyl)taurine $HO_2C-CH(CH_3)-NH-CH_2-CH_2-SO_3H$	Red algae	111
212	Guanidinotaurine $H_2N-C(=NH)-NH-CH_2-CH_2-SO_3H$	Arenicola marina	186
213	N-Methyltaurine $HN-CH_2-CH_2-SO_3H$ $\quad\mid$ $\quad CH_3$	Gelidium cartilagineum	120
214a	N-Dimethyltaurine $H_3C-N-CH_2-CH_2-SO_3H$ $\qquad\mid$ $\qquad CH_3$	Gelidium cartilagineum	120
214b	Mixed disulfide of S-(2-carboxy-3-mercaptopropyl) cysteine and 3-mercaptoisobutyric acid $H_3C-CH-CH_2-S-S-CH_2-CH-CH_2-S-CH_2-CH(NH_2)CO_2H$ $\qquad\mid\qquad\qquad\qquad\qquad\mid$ $\qquad CO_2H\qquad\qquad\qquad CO_2H$	Asparagus officinalis	102c
214c	Mixed disulfide of S-(2-carboxy-3-mercaptopropyl)cysteine $H_2OC-CH(NH_2)-CH_2-S-CH_2-CH-CH_2-S-CH_2-CH-CH_2-S-CH_2-CH(NH_2)CO_2H$ $\qquad\qquad\qquad\qquad\qquad\qquad\mid\qquad\qquad\qquad\mid$ $\qquad\qquad\qquad\qquad\qquad\qquad CO_2H\qquad\qquad CO_2H$	Asparagus officinalis	102c
Group XI: Selenium-containing compounds			
215	Selenocystathionine $H_2OC-CH(NH_2)-CH_2-Se-CH_2-CH_2-$ $\qquad CH(NH_2)CO_2H$	Neptunia amplexicaulis Morinda reticulata Lecythis ollaria	148 149 103
216	Se-Methylselenocysteine $H_3C-Se-CH_2-CH(NH_2)CO_2H$	Astragalus bisulcatus Oonopsis condensata	142 171

217	Selenocysteine-seleninic acid $HO_2Se—CH_2—CH(NH_2)CO_2H$	*Trifolium pratense* *Trifolium repens* *Lolium perenne*	147 147 147
218	Se-Propen-1-yl-selenocysteine selenoxide $H_3C—CH=CH—Se—CH_2—CH(NH_2)CO_2H$ ↑O	*Allium cepa*	173
219	Selenocystine $CH_2—Se—Se—CH_2$ $CH(NH_2)CO_2H \quad CH(NH_2)CO_2H$	*Astragalus pectinatus*	86
220	Selenohomocystine $HO_2C—CH(NH_2)—(CH_2)_2—Se—Se—(CH_2)_2—CH(NH_2)CO_2H$	*Astragalus crotalariae*	211
221	Selenomethionine $H_3C—Se—CH_2—CH_2—CH(NH_2)CO_2H$	*Triticum vulgare*	147
222	Selenomethionine selenoxide $H_3C—Se—(CH_2)_2—CH(NH_2)CO_2H$ O	*Trifolium pratense* *Trifolium repens* *Lolium perenne*	147 147 147
223	Se-Methylselenomethionine $H_3C—Se—CH_2—CH_2—CH(NH_2)CO_2H$ CH_3	*Astragalus*	170

Group XII: Cyclopropyl ring-containing compounds*

224	1-Aminocyclopropane-1-carboxylic acid 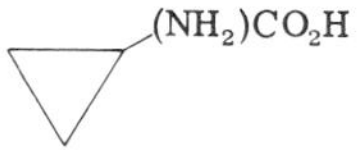 	Rosacea	18 195

* See also 158b and 159.

(continued)

APPENDIX (*Continued*)

No.	Compound	Source	Reference
225	*cis*-2-(2-Carboxycyclopropyl)glycine	*Aesculus parviflora*	60
		Aesculus glabra	54
		trans: Blighia sapida	60
		Aesculus parviflora	60
	H H HO$_2$C CH(NH$_2$)CO$_2$H		
226	*trans*-2-(2-Carboxymethylcyclopropyl)glycine 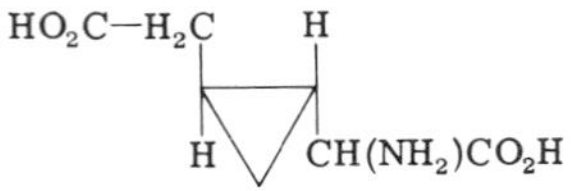	*Blighia unijugata*	60
	HO$_2$C—H$_2$C H H CH(NH$_2$)CO$_2$H		
227	2-(2-Methylenecyclopropyl)glycine 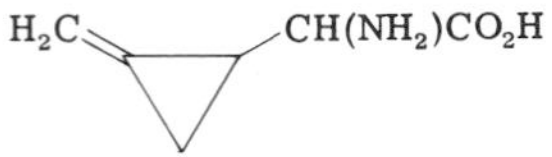	*Litchi chinesis*	75
		Billia hippocastanum	38
		Acer pseudoplatanus	59b
	H$_2$C CH(NH$_2$)CO$_2$H		
228	Hypoglycin[3-(2-methylenecyclopropyl)alanine] 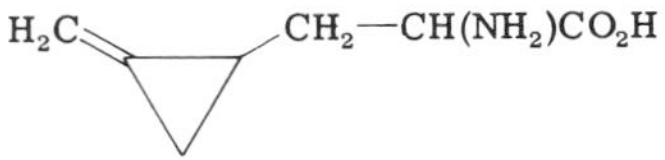	*Blighia sapida*	37
		Billia hippocastanum	38
	H$_2$C CH$_2$—CH(NH$_2$)CO$_2$H		
229	3-Methylhypoglycin 3-(2-methylenecyclopropyl)-3-methylalanine 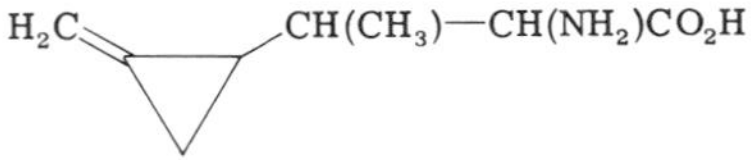	*Aesculus californica*	58
	H$_2$C CH(CH$_3$)—CH(NH$_2$)CO$_2$H		

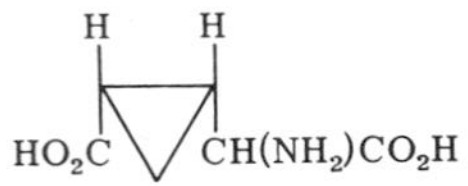

242

ABBREVIATED REFERENCES TO THE APPENDIX

1. Ackermann and Menssen (1960). *Z. Physiol. Chem.* **318**, 212.
2. Andrea and Good (1955). *Plant Physiol.* **30**, 380.
3. Asen *et al.* (1959). *J. Biol. Chem.* **234**, 343.
4. Bell (1961). *Biochim. Biophys. Acta* **47**, 607.
5. Bell (1962). *Biochem. J.* **85**, 91.
6. Bell (1963). *Nature (London)* **199**, 70.
7. Bell and Fellows (1966). *Nature (London)* **210**, 529.
8. Bell and O'Donovan (1966). *Phytochemistry* **5**, 1211.
9. Bell and Tirimanna (1963). *Nature (London)* **197**, 901.
10. Bell and Tirimanna (1964). *Biochem. J.* **91**, 356.
11a. Bell and Tirimanna (1965). *Biochem. J.* **97**, 104.
11b. Bell, Meier, and Sørensen (1981). *Phytochemistry* **9**, 2213.
12a. Bell *et al.* (1971). *Phytochemistry* **10**, 2191.
12b. Bell *et al.* (1980). *J. Am. Chem. Soc.* **102**, 1409.
13. Blake and Fowden (1964). *Biochem. J.* **92**, 136.
14. Brandner and Virtanen (1963). *Acta Chem. Scand.* **17**, 2563.
15. Brewbaker and Hylin (1965). *Crop Sci.* **5**, 348.
16. Brieskorn and Glasz (1964). *Naturwissenschaften* **51**, 216.
17. Brown and Mangat (1969). *Biochim. Biophys. Acta* **177**, 427.
18. Burroughs (1957). *Nature (London)* **179**, 360.
19. Butler, Hegarty, and Peterson (1973). *In* "Chemistry and Biochemistry of Herbage" Vol. I, pp. 1-62. Academic Press, New York.
20a. Buziassy and Mazelis (1974). *Biochim. Biophys. Acta* **86**, 185.
20b. Campos *et al.* (personal communication).
21. Challenger and Hayward (1954). *Biochem. J.* **58**, 10.
22. Clark-Lewis and Mortimer (1959). *Nature (London)* **184**, 1234.
23. Couchman *et al.* (1973). *Phtyochemistry* **12**, 707.
24. Cramer and Spener (1977). *Eur. J. Biochem.* **74**, 495.
25. Cummings and Hudgins (1958). *Am. J. Med. Sci.* **236**, 311.
26a. Daigo (1959). *J. Pharm. Soc. Jn.* **79**, 353, 356.
26b. Daley and Bidwell (1977). *Plant Physiol.* **60**, 109.
27. Dardenne (1970). *Phytochemistry* **9**, 924.
28. Dardenne (1975). *Phytochemistry* **14**, 860.
29a. Dardenne, Marlier, and Casimir (1972). *Phytochemistry* **11**, 2567.
29b. Dardenne, Casimir, and Sørensen (1974). *Phytochemistry* **13**, 2195.
30. Datko, Giovanelli, and Mudd (1974). *J. Biol. Chem.* **249**, 1139.
31. Daxenbichler *et al.* (1972). *Tetrahedron Lett.* **18**, 1801.
32. Dietrichs and Funke (1967). *Holzforschung* **21**, 102.
33. Done and Fowden (1952). *Biochem. J.* **51**, 451.
34. Dunnill and Fowden (1965). *Phytochemistry* **4**, 933.

35. Durham and Milligan (1962). *Biochem. Biophys. Res. Commun.* **7**, 342.
36. Eagles *et al.* (1971). *Biochem. J.* **121**, 425.
37. Ellington *et al.* (1959). *J. Chem. Soc.* p. 80.
38. Eloff and Fowden (1970). *Phytochemistry* **9**, 2423.
39. Eloff and Grobbelaar (1967). *J. S. Afr. Chem. Inst.* **20**, 190.
40. Eloff and Grobbelaar (1969). *Phytochemistry* **8**, 2201.
41. Emery (1965). *Biochemistry* **4**, 1410.
42. Evans, Qureshi, and Bell (1977). *Phytochemistry* **16**, 565.
43. Fang *et al.* (1961). *Sci. Sinica* **10**, 845.
44. Fellows, Hider, and Bell (1977). *Phytochemistry* **16**, 1957.
45. Fellows *et al.* (1979). *Phytochemistry* **18**, 1333.
46. Fowden (1955). *Nature (London)* **176**, 347.
47. Fowden (1958). *Biol. Rev.* **33**, 393.
48. Fowden (1959). *Plant Physiol.* **12**, 657.
49. Fowden (1961). *Biochem. J.* **81**, 154.
50. Fowden (1964). *Annu. Rev. Biochem.* **33**, 173.
51. Fowden (1966). *Biochem. J.* **98**, 57.
52. Fowden (1966). *Nature (London)* **209**, 807.
53. Fowden (1972). *Phytochemistry* **11**, 2271.
54. Fowden (unpublished observation).
55. Fowden and Bryant (1958). *Biochem. J.* **70**, 626.
56. Fowden and Done (1952). *Biochem. J.* **53**.
57. Fowden and Done (1953). *Nature (London)* **171**, 1068.
58. Fowden and Smith (1968). *Phytochemistry,* **7**, 809.
59a. Fowden, Pratt, and Smith (1973). *Phytochemistry* **12**, 1707.
59b. Fowden, Pratt, and Smith (1974). *Phytochemistry* **11**, 3521.
60. Fowden *et al.* (1969). *Phytochemistry* **8**, 437.
61. Fowden *et al.* (1972). *Phytochemistry* **11**, 1105.
62. Gmelin (1959). *Z. Physiol. Chem.* **316**, 164.
63. Gmelin (1962). *Z. Physiol. Chem.* **327**, 186.
64. Gmelin and Hietala (1960). *Z. Physiol. Chem.* **322**, 278.
65. Gmelin and Larsen (1967). *Biochem. Biophys. Acta* **136**, 572.
66. Gmelin and Virtanen (1961). *Ann. Acad. Sci. Fenn. Ser. A 11* **107**, 3.
67. Gmelin and Virtanen (1962). *Suom. Kemistil.* **B35**, 34.
68. Gmelin, Hassenmaier, and Strauss (1957). *Z. Naturforsch.* **126**, 687.
69. Gmelin, Kjaer, and Larsen (1962). *Phytochemistry* **1**, 233.
70. Gmelin, Strauss, and Hasenmaier (1958). *Z. Naturforsch.* **13B**, 252.
71. Gmelin, Strauss, and Hasenmaier (1959). *Z. Physiol. Chem.* **314**, 28.
72. Goas, Larker, and Goas (1970). *Comp. Rend. Acad. Sci., Ser. D.* **271**, 1368.
73. Goodson and Clewer (1917). *J. Chem. Soc.,* pp. 115, 923.
73b. Gray (1972). *Phytochemistry* **11**, 751.
74. Gray and Fowden (1961). *Nature (London)* **189**, 401.
75. Gray and Fowden (1962). *Biochem. J.* **82**, 385.
76. Gray and Fowden (1962). *Nature London)* **193**. 1285.
77. Grobbelaar and Steward (1953). *J. Am. Chem. Soc.* **75**, 4341.
78. Grobbelaar and Steward (1969). *Phytochemistry* **8**, 553.
79. Grobbelaar, Pollard, and Steward (1955). *Nature (London)* **175**, 703.
80. Grobbelaar, Zacharius, and Steward (1954). *J. Am. Chem. Soc.* **76**, 2912.
81. Grove *et al.* (1971). *Tetrahedron Lett.* **47**, 4477.
82a. Hatanaka and Virtanen (1962). *Acta Chem. Scand.* **16**, 514.
82b. Hatanaka *et al.* (1981). *Phytochemistry* **9**, 2291.

83. Hattori and Komamine (1959). *Nature (London)* **183**, 1116.
84. Hegarty and Pound (1968). *Nature (London)* **217**, 354.
85. Hiller-Bombien (1892). *Arch. Pharm.* **230**, 513.
86. Horn and Jones (1941). *J. Biol. Chem.* **139**, 649.
87. Hoshino (1935). *Ann. Chem.* **520**, 31.
88. Hulme and Arthington (1954). *Nature (London)* **173**, 588.
89. Ichihara *et al.* (1973). *Tetrahedron Lett.* **1**, 37.
90. Impellizzeri *et al.* (1975). *Phytochemistry* **14**, 1549.
91. Impellizzeri *et al.* (1977). *Phytochemistry* **16**, 1601.
92. Inatomi, Inugai, and Murakami (1968). *Chem. Pharm. Bull.* **16**, 2521.
93. Inatomi, Inugai, and Murakami (1971). *Chem. Pharm. Bull.* **19**, 216.
94. Inatomi *et al.* (1970). *Chem. Abstr.* **72**, 1854ly.
95. Ito and Fowden (1972). *Phytochemistry* **11**, 2541.
96. Ito and Hashimoto (1966). *Nature (London)* **211**, 417.
97. Jadot and Casimir (1961). *Biochim. Biophys. Acta* **48**, 400.
98. Jadot, Casimir, and Alderweireldt (1963). *Biochim. Biophys. Acta* **78**, 500.
99. Jadot, Casimir, and Loffet (1967). *Biochim. Biophys. Acta* **136**, 79.
100. Jahns (1885). *Chem. Ber.* **18**, 2518.
101. Jahns (1891). *Chem. Ber.* **24**, 2615.
102a. Johns, Lamberton, and Sioumis (1971). *Aust. J. Chem.* **24**, 439.
102b. Kapadia *et al.* (1970). *J. Am. Chem. Soc.* **92**, 6943.
102c. Kasai, Hirakuri, and Sakamura (1981). *Phytochemistry* **9**, 2209.
103. Kerdel-Vegas *et al.* (1965). *Nature (London)* **205**, 1185.
104. King, King, and Warwick (1950). *J. Chem. Soc.* p. 3590.
105. Kjaer and Larsen (1963). *Acta Chem. Scand.* **17**, 2397.
106. Kitagawa and Tomiyama (1929). *J. Biochem. (Tokyo)* **11**, 265.
107. Kittredge and Hughes (1964). *Biochemistry* **3**, 991.
108a. Konishi and Takahashi (1966). *Plant Cell Physiol. (Tokyo)* **7**, 171.
108b. Koyama, Tsujizaki, and Sakamura (1973). *Agric. Chem. Biol.* **37**, 2749.
109. Kirstensen and Larsen (1974). *Phytochemistry* **13**, 1791.
110. Kristensen, Larsen, and Sørensen (1974). *Phytochemistry* **13**, 2803.
111. Kuriyama (1961). *Nature (London)* **192**, 969.
112. Kuriyama, Takagi, and Murata (1960). *Bull. Fac. Fish, Hokkaido Univ.* **11**, 58.
113. Lambein and van Parijs (1968). *Biochem. Biophys. Res. Comm.* **32**, 474.
114. Lambein *et al.* (1976). *Heterocycles* **4**, 567.
115. Larsen (1965). *Acta Chem. Scand.* **19**, 1071.
116. Larsen (1967), *Acta Chem. Scand.* **21**, 1592.
117. Larsen (1968). *Acta Chem. Scand.* **22**, 1369.
118. Larsen *et al.* (1973). *Phytochemistry* **12**, 1713.
119. Leistner and Spenser (1973). *J. Am. Chem. Soc.* **95**, 4715.
120. Lindberg (1955). *Acta Chem. Scand.* **9**, 1323.
121. Linko *et al.* (1953). *Acta Chem. Scand.* **7**, 1310.
122. Liss (1962). *Phytochemistry* **1**, 87.
123. Macnicol (1968). *Biochem. J.* **107**, 473.
124. Madgwick *et al.* (1970). *Arch. Biochem. Biophys.* **141**, 766.
124b. Mandava *et al.* (1974). *Phytochemistry* **13**, 2853.
125. Mansford and Raper (1956). *Ann. Bot. (London)* **20**, 287.
126. Marlier, Dardenne, and Casimir (1972). *Phytochemistry* **11**, 2597.
127. Marumo and Hattori (1970). *Planta* **90**, 208.
128. Matikkala and Virtanen (1962). *Acta Chem. Scand.* **16**, 2461.
129. Matikkala and Virtanen (1963). *Acta Chem. Scand.* **17**, 1799.

130. Mbadiwe (1975). *Phytochemistry* **14,** 1351.

131a. McRorie *et al.* (1954). *J. Am. Chem. Soc.* **76,** 115.

131b. Meier, Olsen, and Sørensen (1979). *Phytochemistry* **18,** 1505.

132. Miersch (1967). *Naturwissenschaften* **54,** 169.

133. Miettinen *et al.* (1953). *Suom. Kemistil.* **B26,** 26.

134. Morgan (1964). *Chem. Ind.,* p. 542.

135. Morris and Thompson (1956). *J. Am. Chem. Soc.* **78,** 1605.

136. Morris and Thompson (1961). *Nature (London)* **190,** 718.

137. Mothes *et al.* (1964). *Z. Naturforsch.* **19B,** 1161.

138. Müller and Schütte (1968). *Z. Naturforsch.* **23B,** 659, 491.

139. Murakoshi *et al.* (1974). *Chem. Pharm. Bull.* **22,** 473.

140a. Mwauluka *et al.* (1975). *Biochem. Physiol. Pflanz.* **168,** 15.

140b. Mwauluka *et al.* (1975). *Phytochemistry* **14,** 1657.

141a. Nakajima and Volcani (1970). *Biochem. Biophys. Res. Commun.* **39,** 28.

141b. Nowaz and Sørensen (1977). *Phytochemistry* **15,** 599.

142. Nigam *et al.* (1969). *Phytochemistry* **8,** 1161.

143a. Noe and Fowden (1960). *Biochem. J.* **77,** 543.

143b. Noina, Noguchi, and Tamaki (1971). *Tetrahedrom Lett.* 2017.

144. Noguchi, Sakuma, and Tamaka (1968). *Phytochemistry* **7,** 1861.

145a. Okuda *et al.* (1975). *Phytochemistry* **14,** 2304.

145b. Olesen and Sørensen (1980). *Phytochemistry,* **19,** 1717.

146. Pant and Fales (1974). *Phytochemistry,* **13,** 1626.

147. Peterson and Butler (1962). *Aust. J. Biol. Sci.* **15,** 126.

148. Peterson and Butler (1967). *Nature (London)* **213,** 599.

149. Peterson and Butler (1971). *Aust. J. Biol. Sci.* **24,** 175.

150. Pilbeam and Bell (1979). *Phytochemistry* **18,** 320.

151. Pollard, Sondheimer, and Steward (1958). *Nature (London)* **182,** 1356.

152. Prochazka (1957). *Coll. Czech. Chem. Commun.* **22,** 333, 654.

153. Przyblska and Pawelkiewicz (1965). *Bull. Acad. Polon. Sci. Ser. Biol.* **13,** 327.

154a. Radhakrishnan and Giri (1954). *Biochem. J.* **58,** 57.

154b. Rao, Adiga, and Sarma (1964). *Biochemistry* **3,** 432.

155. Rao, Ramachandran, and Adiga (1963). *Biochemistry* **2,** 298.

156. Renis and Henze (1959). *Ber. Wiss. Biol.* **133,** 30.

157. Renz (1936). *Z. Physiol. Chem.* **244,** 153.

158. Ressler (1962). *J. Biol. Chem.* **237,** 733.

159. Ressler, Redstone, and Erenberg (1961). *Science* **134,** 188.

160. Reuter (1957). *Flora* **145,** 326.

161. Roche and Lufon (1949). *Compt. Rend.* **229,** 481.

162. Rosenthal (1977). *Q. Rev. Biol.* **52,** 155.

163. Sakato (1950). *J. Agric. Chem. Soc. Jn.* **18,** 262.

164a. Schenk and Schütte (1961). *Naturwissenschaften* **48,** 223.

164b. Schenk and Schütte (1963). *Flora* **153,** 426.

165. Schneider (1958). *Biochem. Z.* **330,** 428.

166. Schwartz (1967). *J. Mol. Biol.* **25,** 571.

167. Scott (1954). *Nature* **173,** 1098.

168. Seneviratne and Fowden (1968). *Phytochemistry* **7,** 1039.

169a. Senoh *et al.* (1964). *Tetrahedron Lett.* **46,** 3431, 3439.

169b. Shewry and Fowden (1976). *Phytochemistry* **15,** 1981.

170. Shrift (1964). *Nature (London)* **201,** 1304.

171a. Shift and Virupaksha (1963). *Biochim. Biophys. Acta* **71,** 483.

171b. Smith (1976) **16,** 1293.

172. Spare and Virtanen (1963). *Acta Chem. Scand.* **17**, 641.
173a. Spare and Virtanen (1964). *Acta Chem. Scand.* **18**, 280.
173b. Sørensen (1976). *Phytochemistry* **15**, 1527.
174. Stewart, Thompson, and Dent (1949). *Science* **110**, 439.
175. Stoll and Seebeck (1947). *Experientia* **3**, 114.
176. Sugii (1961). *Chem. Pharm. Bull (Tokyo)* **12**, 1114.
177. Sung and Fowden (1968). *Phytochemistry* **7**, 2061.
178. Sung *et al.* (1969). *Phytochemistry,* **8**, 1227.
179. Suketa, Sugii, and Suzuki (1970). *Chem. Pharm. Bull.* **18**, 249.
180. Synge and Wood (1956). *Biochem. J.* **64**, 252.
181. Takagi and Okumura (1964). *Bull Jpn. Soc. Sci. Fish.* **30**, 837.
182. Takagi, Hsu, and Takemoto (1970). *Yakugaku Zasshi.* **90**, 899.
183. Takemoto and Sai (1965). *Yakugaku Zasshi.* **85**, 33.
184. Takemoto, Daigo and Takagi (1964). *Yakugaku Zasshi,* **84**, 1176.
185. Tanaka *et al.* (1957). *Proc. Jpn. Acad.* **33**, 47.
186. Thoai and Robin (1954). *Biochim. Biophys. Acta* **13**, 533.
187. Thompson (1956). *Nature (London)* **178**, 593.
188. Thompson and Morris (1968). *Arch. Biochem. Biophys.* **125**, 362.
189. Thompson, Morris, and Hunt (1964). *J. Biol. Chem.* **239**, 1122.
190. Thompson *et al.* (1961). *J. Biol. Chem.* **236**, 1183.
191. Tong and Chaikoff (1955). *J. Biol. Chem.* **215**, 473.
192. Torquati (1913). *Arch. Farmacol. Sper. Sci. Affini* **15**, 308.
193. Tschiersch (1962). *Phytochemistry* **1**, 103.
194. Ueno *et al.* (1957). *Proc. Jpn. Acad.* **33**, 53.
195. Vahatalo and Virtanen (1957). *Acta Chem. Scand.* **11**, 741, 747.
196. van Veen and Hyman (1933). *Geneeskd. Tijdschr. Ned. Indie.* **73**, 991.
197. Vega and Bell (1967). *Phytochemistry* **6**, 759.
198. Virtanen and Berg (1954). *Acta Chem. Scand.* **8**, 1085.
199. Virtanen and Berg (1955). *Acta Chem. Scand.* **9**, 553.
200. Virtanen and Ettala (1957). *Acta Chem. Scand.* **11**, 182.
201. Virtanen and Hietala (1955). *Acta Chem. Scand.* **9**, 175.
202. Virtanen and Kari (1954). *Acta Chem. Scand.* **8**, 1290.
203. Virtanen and Kari (1955). *Acta Chem. Scand.* **9**, 170.
204. Virtanen and Linko (1955). *Acta Chem. Scand.* **9**, 551.
205. Virtanen and Matikkala (1958). *Suom. Kemistil.* **B31**, 191.
206. Virtanen and Matikkala (1959). *Acta Chem. Scand.* **13**, 1898.
207. Virtanen and Matikkala (1960). *Suom. Kemistil* **Ser. B34**, 84.
208. Virtanen and Miettinen (1953). *Biochim. Biophys. Acta* **12**, 181.
209. Virtanen, Hatanaka, and Berlin (1962). *Suom. Kemistil.* **B35**, 52.
210. Virtanen, Uksila, and Matikkala (1954). *Acta Chem. Scand.* **8**, 1094.
211. Virupaksha *et al.* (1966). *Biochim. Biophys. Acta* **130**, 45.
212. Von Klambt (1960). *Naturwissenchaften* **47**, 398.
213. Wada (1930). *Biochem. Z.* **224**, 420.
214. Walker (1954). *J. Biol. Chem.* **204**, 139.
215. Watson and Fowden (1973). *Phytochemistry* **12**, 617.
216a. Welter *et al.* (1978). *Phytochemistry* **17**, 131.
216b. Johnson *et al.* (1974). *Proc. Natl. Acad. Sci.* **71**, 536.
217. Wilding and Stahmann (1962). *Phytochemistry* **1**, 241.
218. Windsor (1951). *J. Biol. Chem.* **192**, 607.
219. Zacharius, Pollard, and Steward (1954). *J. Am. Chem. Soc.* **76**, 1961.

Bibliography

Adams, D. O., and Yang, S. F. (1979). *Proc. Natl. Acad. Sci. U.S.A.* **76**, 170.

Adiga, P. R., Rao, S. L. N., and Sarma, P. S. (1963). *Curr. Sci.* **32**, 153.

Adriaens, P., Meesschaert, B., Wuyts, W., Vanderhaeghe, H., and Eyssen, H. (1977). *J. Chromatogr.* **140**, 103.

Airhart, J., Sibiga, S., Sanders, H., and Khairallah, E. A. (1973). *Anal. Biochem.* **53**, 132.

Aldag, R. W., and Young, J. L. (1970). *Planta* **95**, 187.

Allende, C. C., and Allende, J. E. (1964). *J. Biol. Chem.* **239**, 110.

Amico, V., Oriente, G., and Tringali, C. (1976). *J. Chromatogr.* **116**, 439.

Andrews, R. S., and Pridham, J. B. (1965). *Nature (London)* **205**, 1213.

Applebaum, S. W. (1964). *J. Insect Physiol.* **10**, 783.

Applebaum, S. W., and Schlesinger, H. M. (1977). *In* "EUCARPIA/IOBC Working Group Breeding for Resistance to Insects and Mites," p.143. Wageningen, The Netherlands.

Aronow, L., and Kerdel-Vegas, F. (1965). *Nature (London)* **205**, 1185.

Ascaño, A., and Nicholas, D. J. D. (1977). *Phytochemistry* **16**, 889.

Ashworth, T. S., Brown, E. G., and Roberts, F. M. (1972). *Biochem. J.* **129**, 897.

Atfield, G. N., and Morris, C. J. O. R. (1961). *Biochem. J.* **81**, 606.

Atkins, C. A., Pate, J. S., and Sharkey, P. J. (1975). *Plant Physiol.* **56**, 807.

Attias, J., Schlesinger, M. J., and Schlesinger, S. (1969). *J. Biol. Chem.* **244**, 3810.

Barber, J. T., and Boulter, D. (1963). *Nature (London)* **197**, 1112.

Barrow, M. V., Simpson, C. F., and Miller, E. J. (1974). *Q. Rev. Biol.* **49**, 101.

Beale, S. I. (1978). *Annu. Rev. Plant Physiol.* **29**, 95.

Beale, S. I., and Castelfranco, P. A. (1974). *Plant Physiol.* **53**, 291.

Beeler, T., and Churchich, J. E. (1976). *J. Biol. Chem.* **251**, 5267.

Bell, E. A. (1961). *Arch. Biochem. Biophys.* **47**, 602.

Bell, E. A. (1962a). *Biochem. J.* **83**, 225.

Bell, E. A. (1962b). *Nature (London)* **193**, 1078.

Bell, E. A. (1962c). *Biochem. J.* **85**, 91.

Bell, E. A. (1963a). *Nature (London)* **199**, 70.

Bell, E. A. (1963b). *Biochem. J.* **91**, 358.

Bell, E. A. (1964). *Nature (London)* **203**, 378.

Bell, E. A. (1968). *Nature (London)* **218**, 197.

Bell, E. A. (1978). *In* "Biochemical Aspects of Plant and Animal Coevolution" (J. B. Harborne, ed.), p. 143. Academic Press, New York.

Bell, E. A., and O'Donovan, J. P. (1966). *Phytochemistry* **5**, 1211.

Bell, E. A., and Janzen, D. H. (1971). *Nature (London)* **229**, 136.

Bell, E. A., and Przybylska, J. (1965). *Biochem. J.* **94**, 35P.

Bell, E. A., and Tirimanna, A. S. L. (1963). *Nature (London)* **197,** 901.

Bell, E. A., and Tirimanna, A. S. L. (1964). *Biochem. J.* **91,** 356.

Bell, E. A., and Tirimanna, A. S. L. (1965). *Biochem. J.* **97,** 104.

Bell, E. A., Nulu, J. R., and Cone, C. (1971). *Phytochemistry* **10,** 2191.

Bell, E. A., Lackey, J. A., and Pohill, R. M. (1978). *Biochem. Syst. Ecol.* **6,** 201.

Benson, J. R., and Hare, P. E. (1975). *Proc. Natl. Acad. Sci. U.S.A.* **72,** 619.

Bindon, B. M., and Lamond, D. R. (1966). *Proc. Aust. Soc. Anim. Prod.* **6,** 109.

Blumenthal, S. G., Hendrickson, H. R., Abrol, Y. P., and Conn, E. E. (1968). *J. Biol. Chem.* **243,** 5302.

Blumenthal-Goldschmidt, S., Butler, G. W., and Conn, E. E. (1963). *Nature (London)* **197,** 718.

Boller, T., Herner, R. C., and Kende, H. (1979). *Planta* **145,** 293.

Boyle, J. E., and Fowden, L. (1971). *Phytochemistry* **10,** 2671.

Brendel, K., Corredor, C., and Bressler, R. (1969). *Biochem. Biophys. Res. Commun.* **34,** 340.

Brenner, M., and Neiderwieser, A. (1960). *Experientia* **16,** 378.

Bressler, R., Corredor, C., and Brendel, K. (1969). *Pharmacol. Rev.* **21,** 105.

Brewbaker, J. L., and Hylin, J. W. (1965). *Crop. Sci.* **5,** 348.

Brooks, S. E. H., and Audretsch, J. J. (1976). *In* "Hypoglycin" (E. A. Kean, ed.), p. 57. Academic Press, New York.

Brown, D. H., and Fowden, L. (1966). *Phytochemistry* **5,** 881.

Brown, E. G., and Al-Baldawi, N. F. (1977). *Biochem. J.* **164,** 589.

Brown, E. G., and Mangat, B. S. (1969). *Biochim. Biophys. Acta* **177,** 427.

Brown, E. G., and Silver, A. V. (1966). *Biochim. Biophys. Acta* **119,** 1.

Bryan, J. K. (1980). *In* "The Biochemistry of Plants" (B. J. Miflin, ed.), Vol. 5, pp. 463–452. Academic Press, New York.

Brysk, M. M., Corpe, W. A., and Hankes, L. V. (1969). *J. Bacteriol.* **97,** 322.

Budĕšínský, M., Budzikiewicz, H., Procházka, Z., Ripperger, H., Römer, A., Scholz, G., and Schreiber, K. (1980). *Phytochemistry* **19,** 2295.

Burnell, J. N., and Shrift, A. (1977). *Plant Physiol.* **60,** 670.

Burroughs, L. F. (1960). *J. Sci. Food Agric.* **11,** 14.

Cahn, R. S., Ingold, C. K., and Prelog, V. (1956). *Experientia* **12,** 81.

Carnegie, P. R. (1963). *Biochem. J.* **89,** 459.

Castric, P. A., Farnden, K. J. F., and Conn, E. E. (1972). *Arch. Biochem. Biophys.* **152,** 62.

Chang, L. T. (1960). *J. Formosan Med. Assoc.* **59,** 882.

Charlwood, B. V., and Bell, E. A. (1977). *J. Chromatogr.* **135,** 377.

Cheema, P. S., Padmanaban, G., and Sarma, P. S. (1969a). *Indian J. Biol. Chem.* **6,** 146.

Cheema, P. S., Malathi, K., Padmanaban, G., and Sarma, P. S. (1969b). *Biochem. J.* **112,** 29.

Cheema, P. S., Padmanaban, G., and Sarma, P. S. (1970). *J. Neurochem.* **17,** 1295.

Chen, D. M., Nigam, S. N., and McConnell, W. B. (1970). *Can. J. Biochem.* **48,** 1278.

Chen, K. K., Anderson, R. C., McCowen, M. C., and Harris, P. N. (1951). *J. Pharmacol. Exp. Ther.* **121,** 272.

Chilton, W. S., Tsou, G., Kirk, L., and Benedict, G. R. (1968). *Tetrahedron Lett.* p. 6283.

Chow, C. M., Nigam, S. N., and McConnell, W. B. (1972). *Biochim. Biophys. Acta* **273,** 91.

Chow, C. M., Nigam, S. N., and McConnell, W. B. (1973). *Can. J. Biochem.* **51,** 489.

Chrispeels, M. J. (1970). *Biochem. Biophys. Res. Commun.* **39**, 732.

Chrispeels, M. J., and Sadava, D. E. (1974). *Soc. Dev. Biol.* **30**, 131.

Christensen, H. N., Riggs, T. R., Fisher, H., and Palatine, I. M. (1952). *J. Biol. Chem.* **198**, 17.

Christie, G. S., Madsen, N. P., and Hegarty, M. P. (1969). *Biochem. Pharmacol.* **18**, 693.

Christie, G. S., deMunk, F. G., Madsen, N. P., and Hegarty, M. P. (1971). *Pathology* **3**, 139.

Christie, G. S., Wilson, M., and Hegarty, M. P. (1975). *J. Pathol.* **117**, 195.

Chwalek, B., and Przybylska, J. (1970). *Bull. Acad. Pol. Sci., Ser. Sci. Biol.* **18**, 603.

Circo, R., and Freeman, B. A. (1963). *Anal. Chem.* **35**, 262.

Clandinin, M. T., and Cossins, E. A. (1974). *Phytochemistry* **13**, 585.

Coch, E. H., and Greene, R. C. (1971). *Biochim. Biophys. Acta* **230**, 223.

Cochran, D. G. (1975). *In* "Insect Biochemistry and Function" (D. J. Candy and B. A. Kilby, eds.), p. 177. Chapman & Hall, London.

Consden, R., Gordon, A. H., and Martin, A. J. P. (1944). *Biochem. J.* **38**, 224.

Corredor, C. P. (1976). *In* "Hypoglycin" (E. A. Kean, ed.), p. 145. Academic Press, New York.

Coulter, J. R., and Hann, C. S. (1971). *In* "New Techniques in Amino Acid, Peptide, and Protein Analysis" (A. Niederwieser and G. Pataki, eds.). Ann Arbor Sci. Publ., Ann Arbor, Michigan.

Cowie, D. B., and Cohen, G. N. (1957). *Biochim. Biophys. Acta* **26**, 252.

Crounse, R. G., Maxwell, J. D., and Blank, H. (1962). *Nature (London)* **194**, 694.

Culvenor, C. C. J., Foster, M. C., and Hegarty, M. P. (1969). *Aust. J. Chem.* **24**, 371.

Cummins, L. M., and Martin, J. L. (1967). *Biochemistry* **10**, 3162.

Dahlman, D. L., and Rosenthal, G. A. (1975). *Comp. Biochem. Physiol. A* **51A**, 33.

Damodaran, M., and Narayanan, K. G. A. (1939). *Biochem. J.* **33**, 1740.

Dardenne, G. A., Bell, E. A., Nulu, J. R., and Cone, C. (1972). *Phytochemistry* **11**, 791.

Dardenne, G. A., Casimir, J., and Sørensen, H. (1974). *Phytochemistry* **13**, 2195.

Dardenne, G. A., Larsen, P. O., and Wieczorkowska, E. (1975). *Biochim. Biophys. Acta* **381**, 416.

Datko, A. H., Giovanelli, J., and Mudd, S. H. (1974). *J. Biol. Chem.* **249**, 1139.

Datko, A. H., Mudd. S. H., and Giovanelli, J. (1977). *J. Biol. Chem.* **252**, 3436.

DaVanzo, J. P., Matthews, R. J., Young, G. A., and Wingerson, F. (1964). *Toxicol. Appl. Pharmacol.* **6**, 396.

Dawson, R. M. C., Elliott, D. C., Elliott, W. H., and Jones, K. M. (1969). "Data for Biochemical Research." Oxford Univ. Press, London and New York.

Daxenbichler, M. E., van Etten, C. H., Hallinan, E. A., and Earle, F. R. (1971). *J. Med. Chem.* **14**, 463.

Dekker, E. E., and Maitra, U. (1962). *J. Biol. Chem.* **237**, 2218.

Dent, C. E., Stepka, W., and Steward, F. C. (1947). *Nature (London)* **160**, 682.

deRenzo, E. C., McKerns, K. W., Bird, H., Cekleniak, W., Coulomb, B. S., and Kaleita, E. (1958). *Biochem. Pharmacol.* **1**, 230.

Despontin, J., Marlier, M., and Dardenne, G. A. (1977). *Phytochemistry* **16**, 387.

Dewreede, S., and Wayman, O. (1970). *Teratology* **3**, 21.

Dodd, W. A., and Cossins, E. A. (1968). *Phytochemistry* **7**, 2143.

Doney, R. C., and Thompson, J. F. (1971). *Phytochemistry* **10**, 1745.

Dossaji, S. F., and Bell, E. A. (1973). *Phytochemistry* **12**, 143.

Dougall, D. K., and Fulton, M. M. (1967). *Plant Physiol.* **42**, 941.

Dunnill, P. M., and Fowden, L. (1965b). *Phytochemistry* **4**, 445.

Dus, K., Bartsch, R. G., and Kamen, M. D. (1962). *J. Biol. Chem.* **237,** 3083.
Ehrlich, P. R., and Raven, P. H. (1965). *Evolution* **18,** 586.
Ellington, E. V. (1976). *In* "Hypoglycin" (E. A. Kean, ed.), p. 1. Academic Press, New York.
Ellis, B. E. (1975). *Phytochemistry* **15,** 489.
Eloff, J. N., and Fowden, L. (1970). *Phytochemistry* **9,** 2423.
Enoch, H. G., and Lester, R. L. (1975). *J. Biol. Chem.* **250,** 6693.
Entman, M., and Bressler, R. (1967). *Mol. Pharmacol.* **3,** 333.
Eustice, D. C., Foster, I., Kull, F. J., and Shrift, A. (1980). *Plant Physiol.* **66,** 182.
Evans, C., Asher, C., and Johnson, C. M. (1968). *Aust. J. Biol. Sci.* **21,** 13.
Evans, C. S., and Bell, E. A. (1979). *Phytochemistry* **18,** 1807.
Evans, C. S., Qureshi, M. Y., and Bell, E. A. (1977). *Phytochemistry* **16,** 565.
Everett, G. M., and Borcherding, J. W. (1970). *Science* **16,** 849.
Eyre, D. R. (1980). *Science* **207,** 1315.
Feeny, P. (1975). *In* "Coevolution of Animals and Plants" (L. E. Gilbert and P. H. Raven, eds.), p. 3. Univ. of Texas Press, Austin.
Fellows, L. E., and Bell, E. A. (1970). *Phytochemistry* **9,** 2389.
Finney, D. J. (1971). "Probit Analysis," 3rd ed., p. 333. Cambridge Univ. Press, London and New York.
Floss, H. G., Hadwiger, L., and Conn, E. E. (1965). *Nature (London)* **208,** 1207.
Fowden, L. (1955). *Biochem. J.* **64,** 323.
Fowden, L. (1958). *Nature (London)* **182,** 406.
Fowden, L. (1960). *J. Exp. Bot.* **11,** 302.
Fowden, L. (1963). *J. Exp. Bot.* **14,** 387.
Fowden, L. (1970). *Prog. Phytochem.* **2,** 203.
Fowden, L. (1972). *Phytochemistry* **11,** 2271.
Fowden, L. (1973). *In* "Biosynthesis and Its Control in Plants" (B. V. Millborrow, ed.), p. 324. Academic Press, New York.
Fowden, L. (1976). *In* "Hypoglycin" (E. A. Kean, ed.), p. 11. Academic Press, New York.
Fowden, L., and Bell, E. A. (1965). *Nature (London)* **206,** 110.
Fowden, L., and Bryant, M. (1959). *Biochem. J.* **71,** 210.
Fowden, L., and Frankton, J. B. (1968). *Phytochemistry* **7,** 1077.
Fowden, L., and Lea, P. J. (1979). *In* "Herbivores: Their Interaction with Secondary Plant Metabolites" (G. A. Rosenthal and D. H. Janzen, eds.), p. 135. Academic Press, New York.
Fowden, L., and Mazelis, M. (1971). *Phytochemistry* **10,** 359.
Fowden, L., and Pratt, H. M. (1973). *Phytochemistry* **12,** 1677.
Fowden, L., and Richmond, M. H. (1963). *Biochim. Biophys. Acta* **71,** 459.
Fowden, L., and Smith, A. (1968). *Phytochemistry* **7,** 809.
Fowden, L., and Steward, F. C. (1957). *Ann. Bot. (London)* [N.S.] **21,** 60.
Fowden, L., Lewis, D., and Tristram, H. (1967). *Adv. Enzymol.* **29,** 89.
Fowden, L., Smith, A., Millington, D. S., and Sheppard, R. C. (1969). *Phytochemistry* **8,** 437.
Fowden, L., Anderson, J. W., and Smith, A. (1970). *Phytochemistry* **9,** 2349.
Fowden, L., MacGibbon, C. M., Mellon, F. A., and Sheppard, R. C. (1972a). *Phytochemistry* **11,** 1105.
Fowden, L., Pratt, H. M., and Smith, A. (1972b). *Phytochemistry* **11,** 3521.
Fowden, L., Lea, P. J., and Bell, E. A. (1979). *Adv. Enzymol.* **50,** 117.
Freeland, W. J., and Janzen, D. H. (1974). *Am. Nat.* **108,** 269.

Frenkel, M. J., Gillespie, J. M., and Reis, P. J. (1975). *Aust. J. Biol. Sci.* **28**, 331.

Frisch, D. M., Dunnill, P. M., Smith, A., and Fowden, L. (1967). *Phytochemistry* **6**, 921.

Fugita, Y. (1959). *Bull. Chem. Soc. Jpn.* **32**, 439.

Fugita, Y. (1961). *J. Biochem. (Tokyo)* **49**, 468.

Fugita, Y., Kollonitsch, J., and Witkop, B. (1965). *J. Am. Chem. Soc.* **87**, 2030.

Fujihara, S., and Yamaguchi, M. (1980). *Plant Physiol.* **66**, 139.

Gaetani, G., Salvidio, E., Pannacciulli, I., Ajmar, F., and Paravidino, G. (1970). *Experientia* **26**, 785.

Gilbertson, T. J., (1972). *Phytochemistry* **11**, 1737.

Giovanelli, J. (1966). *Biochim. Biophys. Acta* **118**, 124.

Giovanelli, J., and Mudd, S. H. (1967). *Biochem. Biophys. Res. Commun.* **27**, 150.

Giovanelli, J., and Mudd, S. H. (1968). *Biochem. Biophys. Res. Commun.* **31**, 275.

Giovanelli, J., and Mudd, S. H. (1971). *Biochim. Biophys. Acta* **227**, 654.

Giovanelli, J., Owens, L. D., and Mudd, S. H. (1971). *Biochim. Biophys. Acta* **227**, 671.

Giovanelli, J., Owens, L. D., and Mudd, S. H. (1973). *Plant Physiol.* **51**, 492.

Giovanelli, J., Mudd, S. H., and Datko, A. H. (1974). *Plant Physiol.* **54**, 725.

Giovanelli, J., Mudd, S. H., and Datko, A. H. (1978). *J. Biol. Chem.* **253**, 5665.

Giovanelli, J., Mudd, S. H., and Datko, A. H. (1979). *In* "Third International Meeting of Low Molecular Weight Sulfur-Containing Natural Products" (D. Cavallini *et al.*, eds.), Plenum, New York.

Giovanelli, J., Mudd, S. H., and Datko, A. H. (1980). *In* "The Biochemistry of Plants" (B. J. Miflin, ed.), Vol. 5, p. 453. Academic Press, New York.

Gmelin, R. (1959). *Hoppe-Seyler's Z. Physiol. Chem.* **316**, 164.

Gmelin, R. (1961). *Acta Chem. Scand.* **15**, 1188.

Gmelin, R., Strauss, G., and Hasenmaier, G. (1958). *Z. Naturforsch., B: Anorg. Chem., Org. Chem., Biochem., Biophys., Biol.* **13B**, 252.

Gmelin, R., Strauss, G., and Hasenmaier, G. (1959). *Hoppe-Seyler's Z. Physiol. Chem.* **314**, 28.

Gmelin, R., Kjaer, A., and Larsen, P. O. (1962). *Phytochemistry* **1**, 233.

Godwin, K. O., and Fuss, C. N. (1972). *Aust. J. Biol. Sci.* **25**, 865.

Godwin, K. O., Handreck, K. A., and Fuss, C. N. (1971). *Aust. J. Biol. Sci.* **24**, 1251.

Gonzalez, V., Brewbaker, J. L., and Hamill, D. E. (1967). *Crop Sci.* **7**, 140.

Goore, M. Y., and Thompson, J. F. (1967). *Biochim. Biophys. Acta* **132**, 15.

Gray, D. O., and Fowden, L. (1962). *Biochem. J.* **82**, 385.

Gray, D. O., Blake, J., Brown, D. H., and Fowden, L. (1964). *J. Chromatogr.* **13**, 276.

Greene, R. C., and Davis, N. B. (1960). *Biochim. Biophys. Acta* **43**, 360.

Greenstein, J. F., and Winitz, M. (1961). "Chemistry of the Amino Acids," Vols. 1-3. Wiley, New York.

Griffith, T., and Conn, E. E. (1973). *Phytochemistry* **12**, 1651.

Grove, J. A., Gilbertson, T. J., Hammerstedt, R. H., and Henderson, L. M. (1969). *Biochim. Biophys. Acta* **184**, 329.

Guggenheim, M. (1913). *Hoppe-Seyler's Z. Physiol. Chem.* **88**, 276.

Gulati, D. K., Chambers, C. L., Rosenthal, G. A., and Sabharwal, P. S. (1981). *Envir. Exp. Bot.* **121**, 225.

Gulland, J. M., and Morris, C. J. O. R. (1935). *J. Chem. Soc.* p. 763.

Gupta, R. N., and Spenser, I. D. (1970). *Phytochemistry* **9**, 2329.

Hamilton, J. W. (1975). *J. Agric. Food Chem.* **23**, 1150.

Hamilton, R. I., Donaldson, L. E., and Lambourne, L. J. (1968). *Aust. Vet. J.* **44**, 484.

Hardy, M. J. (1974). *Anal. Biochem.* **57**, 529.

Hasegawa, M., and Matsubara, I. (1975). *Anal. Biochem.* **63,** 308.

Hassall, C. H., and John, D. I. (1960). *J. Chem. Soc. (London)* 4112.

Hassall, C. H., and Reyle, K. (1955). *Biochem. J.* **60,** 334.

Hasse, K., Ratych, O. T., and Salnikow, J. (1967). *Hoppe-Seyler's Z. Physiol. Chem.* **348,** 843.

Hassid, E., Applebaum, S. W., and Birk, Y. (1976). *Phytoparasitica* **4,** 173.

Hattori, S., and Komamine, A. (1959). *Nature (London)* **183,** 1116.

Hegarty, M. P., and Pound, A. W. (1968). *Nature (London)* **217,** 354.

Hegarty, M. P., and Pound, A. W. (1970). *Aust. J. Biol. Sci.* **23,** 831.

Hegarty, M. P., Schinckel, P. G., and Court, R. D. (1964). *Aust. J. Agric. Res.* **15,** 153.

Hegarty, M. P., Court, R. D., Christie, G. S., and Lee, C. P. (1976). *Aust. Vet. J.* **52,** 490.

Hegarty, M. P., Lee, C. P., Christie, G. S., Court, R. D., and Haydock, K. P. (1979). *Aust. J. Biol. Sci.* **32,** 27.

Henke, L. A. (1959). *Proc. Pac. Sci. Congr., 8th, 1953* Vol. IVB, p. 591.

Herrmann, R. L., Lou, M. F., and White, C. W. (1966). *Biochim. Biophys. Acta* **121,** 79.

Hider, R. C., and John, D. I. (1973). *Phytochemistry* **12,** 119.

Hijman, A. J., and van Veen, A. G. (1936). *Geneeskd. Tijdschr. Ned.-Indie* **76,** 840.

Hill, K. R. (1952). *West Indian Med. J.* **1,** 243.

Hornykiewicz, O. (1966). *Pharmacol. Rev.* **18,** 925.

Huber, R. E., and Criddle, R. S. (1967). *Arch. Biochem. Biophys.* **122,** 164.

Hunt, G. E., and Thompson, J. F. (1971). *Biochem. Prep.* **13,** p. 41.

Hurych, J., Chrapil, M., Tichy, M., and Beniae, F. (1967). *Eur. J. Biochem.* **3,** 31.

Hutton, E. M., and Gray, S. G. (1959). *Eur. J. Exp. Agric.* **27,** 187.

Hutton, E. M., Windrum, G. M., and Kratzing, C. C. (1958a). *J. Nutr.* **64,** 321.

Hutton, E. M., Windrum, G. M., and Kratzing, C. C. (1958b). *J. Nutr.* **65,** 429.

Hylin, J. W. (1964). *Phytochemistry* **3,** 161.

Hylin, J. W. (1969). *J. Agric. Food Chem.* **17,** 492.

Hylin, J. W., and Lichton, I. J. (1965). *Biochem. Pharmacol.* **14,** 1167.

Inatomi, H., Inugai, F., and Murakomi, T. (1968). *Chem. Pharm. Bull.* **16,** 2521.

Inatomi, K., and Slaughter, J. C. (1970). *J. Exp. Bot.* **72,** 561.

Iwase, H., and Murai, A. (1977). *Anal. Biochem.* **78,** 340.

Janzen, D. H. (1969). *Evolution* **23,** 1.

Janzen, D. H. (1979). In "Herbivores: Their Interaction with Secondary Plant Metabolites" (G. A. Rosenthal and D. H. Janzen, eds.), p. 331. Academic Press, New York.

Janzen, D. H., Juster, H. B., and Bell, E. A. (1977). *Phytochemistry* **16,** 223.

Jenkins, K. J. (1968). *Can. J. Biochem.* **46,** 1417.

Jensen, R. A., and Pierson, D. L. (1975). *Nature (London)* **254,** 667.

Johnston, G. A. R. (1973). In "Neuropoisons: Their Pathophysiological Actions" (L. L. Simpson and D. R. Curtis, eds.), p. 179. Plenum, New York.

Johnston, G. A. R., and Lloyd, H. J. (1967). *Aust. J. Biol. Sci.* **20,** 1241.

Johnstone, J. H. (1956). *Biochem. J.* **64,** 21.

Jones, R. J., Blunt, C. G., and Holmes, J. H. G. (1976). *Trop. Grasslands* **10,** 113.

Joshi, H. S. (1968). *Aust. J. Agric. Res.* **19,** 341.

Joy, K. W. (1969). *Plant Physiol.* **44,** 849.

Kammer, A., Dahlman, D. L., and Rosenthal, G. A. (1978). *J. Exp. Biol.* **75,** 123.

Kasai, T., and Larsen, P. O. (1980). *Prog. Chem. Org. Nat. Prod.,* **39,** 173.

Kasting, R., and Delwiche, C. C. (1957). *Plant Physiol.* **32,** 471.

Kasting, R., and Delwiche, C. C. (1958). *Plant Physiol.* **33,** 350.

Katunuma, N., Okada, M., Matsuzawa, T., and Otsuka, Y. (1965). *J. Biochem. (Tokyo)* **57,** 445.

Kawerau, E., and Wieland, T. (1951). *Nature (London)* **168,** 77.

Kean, E. A., and Hare, E. R. (1980). *Phytochemistry* **19,** 199.

Kekomäki, M., Rahiala, E.-L., and Räihä, N. C. R. (1969). *Ann. Med. Exp. Fenn.* **47,** 33.

Kerdel-Vegas, F. (1964). *J. Invest. Dermatol.* **42,** 91.

Kerdel-Vegas, F., Wagner, F., Russell, P. B., Grant, N. H., Alburn, H. E., Clark, D. E., and Miller, J. A. (1965). *Nature (London)* **205,** 1186.

Kitagawa, M., and Tomiyama, T. (1929). *J. Biochem. (Tokyo)* **25,** 23.

Kitagawa, M., and Yamada, H. (1932). *J. Biochem. (Tokyo)* **16,** 339.

Kizer, J. S., Nemeroff, C. B., and Youngblood, W. W. (1978). *Pharmacol. Rev.* **29,** 301.

Kjaer, A., and Larsen, P. O. (1974). "Biosynthesis," Vol. 4. Chemical Society, London.

Kjaer, A., Larsen, O. P., and Gmelin, R. (1959). *Experientia* **15,** 253.

Konze, J. R., and Kende, H. (1979). *Plant Physiol.* **63,** 507.

Kosower, N. S., and Kosower, E. M. (1967). *Nature (London)* **215,** 285.

Kraneveld, and Djaenoedin. (1950). *Hemera Zoa* **57,** 623.

Krauss, G. Y., and Reinbothe, H. (1973). *Phytochemistry* **12,** 125.

Kristensen, I., and Larsen, P. O. (1974). *Phytochemistry* **13,** 2791.

Kruse, P. F., White, P. B., Carter, H. A., and McCoy, T. A. (1959). *Cancer Res.* **19,** 122.

Kuttan, R., and Radhakrishnan, A. N. (1970). *Biochem. J.* **117,** 1015.

Lackey, J. A. (1977). *Bot. J. Linn. Soc.* **74,** 163.

Laghai, A., and Jordan, P. M. (1977). *Biochem. Soc. Trans.* **5,** 299.

Lallier, R. (1965). *Exp. Cell Res.* **40,** 630.

Lambein, F., and van Parijs, R. (1968). *Biochem. Biophys. Res. Commun.* **32,** 474.

Lambein, F., and van Parijs, R. (1970). *Biochem. Biophys. Res. Commun.* **40,** 557.

Lambein, F., and van Parijs, R. (1974). *Biochem. Biophys. Res. Commun.* **61,** 155.

Lambein, F., Kuo, Y.-H., and van Parijs, R. (1976). *Heterocycles* **4,** 567.

Lamothe, P. J., and McCormick, P. G. (1973). *Anal. Chem.* **45,** 1906.

Landau, A. J., Fuerst, R., and Awapara, J. (1951). *Anal. Chem.* **23,** 162.

Lane, J. M., Dehm, P., and Prockop, D. J. (1971). *Biochim. Biophys. Acta* **236,** 51.

Larsen, P. O., (1967). *Biochim. Biophys. Acta* **141,** 27.

Larsen, P. O., and Norris, F. (1976). *Phytochemistry* **15,** 1761.

Larsen, P. O., and Sørensen, H. (1968). *Biochim. Biophys. Acta* **156,** 190.

Larsen, P. O., and Wieczorkowska, E., (1975). *Biochim. Biophys. Acta* **381,** 409.

Larsen, P. O., Sørensen, H., Cochran, D. W., Hagaman, E. W., and Wenkert, E. (1973). *Phytochemistry* **12,** 1713.

Larsen, P. O., Onderka, D. K., and Floss, H. G. (1975). *Biochim. Biophys. Acta* **381,** 397.

Larsen, P. O., Sørensen, F. T., and Wieczorkowska, E. (1978). *Phytochemistry* **17,** 549.

LaRue, T. A., and Child, J. J. (1975). *Phytochemistry* **14,** 2512.

Lea, P. J., and Fowden, L. (1972). *Phytochemistry* **11,** 2129.

Lea, P. J., and Fowden, L. (1973). *Phytochemistry* **12,** 1903.

Lea, P. J., and Fowden, L. (1975a). *Biochem. Physiol. Pflanz.* **168,** 3.

Lea, P. J., and Fowden, L. (1975b). *Proc. R. Soc. London, Ser. B* **192,** 13.

Lea, P. J., and Miflin, B. J. (1974). *Nature (London)* **251,** 614.

Lea, P. J., and Norris, R. D. (1976). *Phytochemistry* **15,** 585.

Lea, P. J., Hughes, J. S., and Miflin, B. J. (1979). *J. Exp. Bot.* **30,** 529.

Leete, E., (1964). *J. Am. Chem. Soc.* **86,** 3162.

Leete, E., Davis, G. E., Hutchinson, C. R., Woo, K. W., and Chedekel, M. R. (1974). *Phytochemistry* **13,** 427.

Leistner, E., Gupta, R. N., and Spencer, I. P. (1973). *J. Am. Chem. Soc.* **95,** 4040.

Lever, M., and Butler, G. W. (1971a). *J. Exp. Bot.* **22,** 279.

Lever, M., and Butler, G. W. (1971b). *J. Exp. Bot.* **22,** 285.

Levy, A. L., and Chung, D. (1953). *Anal. Chem.* **25,** 396.

Lewis, B. G. (1976). *In* "Environmental Biogeochemistry" (J. O. Nriagu, ed.), p. 389. Ann Arbor Sci. Publ., Ann Arbor, Michigan.

Lewis, B. G., Johnson, C. M., and Broyer, T. C. (1971). *Biochim. Biophys. Acta* **237,** 603.

Lewis, H. B., and Schulert, A. R. (1949). *Proc. Soc. Exp. Biol. Med.* **71,** 440.

Lewis, H. B., Fajans, R. S., Esterer, M. B., Shen, C.-W., and Oliphant, M. (1948). *J. Nutr.* **36,** 537.

Lin, J.-Y., and Ling, K.-H. (1961). *J. Formosan Med. Assoc.* **60,** 657.

Lin, J.-Y., Shih, Y.-M., and Ling, K.-H. (1962a). *J. Formosan Med. Assoc.* **61,** 997.

Lin, J.-Y., Lin, K.-T., and Ling, K.-H. (1963). *J. Formosan Med. Assoc.* **62,** 587.

Lin, K.-T., Lin, J.-K., and Tung, T.-C. (1964). *J. Formosan Med. Assoc.* **63,** 10.

Lindeberg, E. G. G. (1976). *J. Chromatogr.* **117,** 439.

Linko, P., and Virtanen, A. I. (1958). *Acta Chem. Scand.* **12,** 68.

Liss, I. (1961). *Flora (Jena, 1818 –1965)* **151,** 351.

Lloyd, N. D. H., and Joy, K. W. (1978). *Biochem. Biophys. Res. Commun.* **81,** 186.

Maas, W. K. (1960). *Cold Spring Harbor Symp. Quant. Biol.* **26,** 183.

McCaldin, D. J. (1960). *Chem. Rev.* **60,** 39.

McFarren, E. F. (1951). *Anal. Chem.* **23,** 168.

McKay, G. F., Lalich, J. J., Schilling, E. D., and Strong, F. M. (1954). *Arch. Biochem. Biophys.* **52,** 313.

McKerns, K. W., Bird, H. H., Kaleita, E., Coulomb, B. S., and deRenzo, E. C. (1960). *Biochem. Pharmacol.* **3,** 305.

McKey, D. (1979). *In* "Herbivores: Their Interaction with Secondary Plant Metabolites" (G. A. Rosenthal and D. H. Janzen, eds.), p. 56. Academic Press, New York.

McMahon, D., and Langstroth, P. (1972). *J. Gen. Microbiol.* **73,** 239.

Madsen, N. P., and Hegarty, M. P. (1970). *Biochem. Pharmacol.* **19,** 2391.

Madsen, N. P., Christie, G. S., and Hegarty, M. P. (1970). *Biochem. Pharmacol.* **19,** 853.

Mae, T., Ohira, K., and Fujiwara, A. (1971). *Plant Cell Physiol.* **12,** 1.

Maeda, M., and Tsuji, A. (1973). *Anal. Biochem.* **52,** 555.

Makisumi, S. (1961). *J. Biochem. (Tokyo)* **49,** 284.

Makita, M., Yamamoto, S., Sakai, K., and Shiraishi, M. (1976). *J. Chromatogr.* **124,** 92.

Malathi, K., Padmanaban, G., Rao, S. L. N., and Sarma, P. S. (1967). *Biochim. Biophys. Acta* **141,** 71.

Malathi, K., Padmanaban, B., and Sarma, P. S. (1970). *Phytochemistry* **9,** 1603.

Martin, A. J. P., and Synge, R. L. M. (1941). *Biochem. J.* **35,** 1358.

Martin, J. L., Shrift, A., and Gerlach, M. L. (1971). *Phytochemistry* **10,** 945.

Masada, M., Fukushima, K., and Tamura, G. (1975). *J. Biochem. (Tokyo)* **77,** 1107.

Mason, M. M., and Whiting, M. G. (1966). *Fed. Proc., Fed. Am. Soc. Exp. Biol.* **25,** 533.

Matsumoto, H., Smith, E. G., and Sherman, G. D. (1951). *Arch. Biochem. Biophys.* **33,** 201.

Mazelis, M. (1963). *Phytochemistry* **2,** 15.

Mazelis, M., and Creveling, R. K. (1978). *Plant Physiol.* **62,** 798.

Mazelis, M., and Fowden, L. (1972). *Phytochemistry* **11,** 619.

Mazelis, M., Whatley, F. R., and Whatley, J. (1977). *FEBS Lett.* **84,** 236.

Mead, R. J., and Segal, W. (1973). *Phytochemistry* **12,** 1977.

Meister, A. (1973). *Science* **180,** 33.

Meister, A., Radhakrishnan, A. N., and Buckley, S. D. (1957). *J. Biol. Chem.* **229,** 789.

Menske, R. H. F. (1937). *Can. J. Res. Sec. B* **15,** 84.

Mestichelli, L. J. J., Gupta, R. N., and Spenser, I. D. (1979). *J. Biol. Chem.* **254,** 640.

Miersch, J. (1967). *Naturwissenschaften* **54,** 169.

Miflin, B. J., and Lea, P. J. (1977). *Annu. Rev. Plant Physiol.* **28,** 299.

Miflin, B. J., and Lea, P. J. (1980). *In* "The Biochemistry of Plants" A Comprehensive Treatise," (B. J. Miflin, ed.), Vol. 5, p. 169. Academic Press, New York.

Mikeš, O. (1966). "Laboratory Handbook of Chromatographic Methods." Van Nostrand-Reinhold, Princeton, New Jersey.

Miller, R. W., and Smith, C. R., Jr. (1973). *J. Agric. Food Chem.* **21,** 909.

Mitchell, D. J., and Bidwell, R. G. S. (1970). *Can. J. Bot.* **48,** 2001.

Mitra, S. K., and Mehler, A. H. (1967). *J. Biol. Chem.* **242,** 5490.

Moffat, E. D., and Lytle, R. I. (1959). *Anal. Chem.* **31,** 926.

Møller, B. L. (1974). *Plant Physiol.* **54,** 638.

Møller, B. L. (1976). *Plant Physiol.* **57,** 687.

Moore, S., and Stein, W. H. (1949). *J. Biol. Chem.* **178,** 53.

Moore, S., and Stein, W. H. (1954). *J. Biol. Chem.* **211,** 893.

Morris, C. J., and Morris, P. (1973). "Separation Methods in Biochemistry," 2nd ed. Halsted Press, New York.

Morris, C. J., and Thompson, J. F. (1964). *Arch. Biochem. Biophys.* **119,** 269.

Morris, C. J., and Thompson, J. F. (1965). *Arch. Biochem. Biophys.* **110,** 506.

Morris, C. J., and Thompson, J. F. (1975). *Plant Physiol.* **55,** 960.

Morris, C. J., and Thompson, J. F. (1977). *Plant Physiol.* **59,** 684.

Mudd, S. H. (1960). *Biochim. Biophys. Acta* **38,** 354.

Müller, P., and Schütte, H. R. (1968). *Z. Naturforsch., B: Anorg. Chem., Org. Chem. Biochem. Biophys. Biol.* **23B,** 491.

Müller, P. and Schütte, H. R. (1971). *Biochem. Physiol. Pflanz.* **162,** 234.

Murakoshi, I., Kuramoto, H., Haginiwa, J., and Fowden, L. (1972a). *Phytochemistry* **11,** 177.

Murakoshi, I., Kato, F., Haginiwa, H., and Fowden, L. (1972b). *Chem. Pharm. Bull.* **21,** 918.

Murakoshi, I., Kato, F., Haginiwa, J., and Takemoto, T., (1974). *Chem. Pharm. Bull.* **22,** 473.

Murakoshi, I., Ikegami, F., Ookawa, N., Ariki, T., Haginiwa, J., Kuo, Y.-H., and Lambein, F. (1978). *Phytochemistry* **17,** 1571.

Murr, D. P., and Yang, S. F. (1975). *Plant Physiol.* **55,** 79.

Mwauluka, K., Charlwood, B. V., Briggs, J. M., and Bell, E. A. (1975). *Biochem. Physiol. Pflanz.* **168,** 15.

Nagarajan, V., Mohan, V. S., and Gopalan, C. (1965). *Indian J. Med. Res.* **53,** 269.

Nagasawa, T., Takagi, H., Kawakami, K., Suzuki, T., and Sahashi, Y. (1961). *Agric. Biol. Chem.* **25,** 441.

Narayanan, A. S., Siegel, R. C., and Martin, G. R. (1972). *Biochem. Biophys. Res. Commun.* **46,** 745.

Natelson, S., Koller, A., Tseng, H.-Y., and Dods, R. F. (1977). *Clin. Chem. (Winston-Salem, N.C.)* **23,** 960.

Navon, A., and Bernays, E. A. (1978). *Comp. Biochem. Biophys. A* **59A,** 161.

Naylor, A. W. (1959). *Symp. Soc. Exp. Biol.* **13,** 193.

Neurath, A. R., Wiener, F. P., Rubin, B. A., and Hartzell, R. W. (1970). *Biochem. Biophys. Res. Commun.* **41,** 1509.

Ng, B. H., and Anderson, J. W. (1978). *Phytochemistry* **17,** 879.

Ngo, T. T., and Shargool, P. D. (1974). *Can. J. Biochem.* **52,** 435.

Nigam, S. N., and McConnell, W. B. (1976). *Biochim. Biophys. Acta* **437,** 116.

Nigam, S. N., and Ressler, C. (1966). *Biochemistry* **5,** 3426.

Nigam, S. N., Tu, J.-I., and McConnell, W. B. (1969). *Phytochemistry* **8,** 1161.

Noma, M., Inoue, N., Kawashima, N., and Noguchi, M., (1978). *Phytochemistry* **17,** 991.

Nordfeldt, S., Henke, L. A., Marita, K., Matsumoto, H., Takahashi, M., Young, O. R., Willers, E. H., and Cross, R. F. (1952). *Tech. Bull.—Hawaii Agric. Exp. Stn.* **15.**

Norris, R. D., and Fowden, L. (1972). *Phytochemistry* **11,** 2921.

Notation, A. D., and Spenser, I. D. (1964). *Can. J. Biochem.* **42,** 1803.

Nowacki, E., and Przybylska, J. (1961). *Bull. Acad. Pol. Sci., Ser. Sci. Biol.* **9,** 277.

Oaks, J., and Johnson, F. J. (1972). *Phytochemistry* **11,** 3465.

Oaks, A., Stulen, I., Jones, K., Winspear, M. J., Misra, S., and Boesel, I. L. (1980). *Planta* **148,** 477.

O'Dell, B. L., Elsden, D. F., Thomas, J., Partridge, S. M., Smith, R. H., and Palmer, R. (1966). *Nature (London)* **209,** 401.

Okuda, T., Yoshida, T., Shiota, N., and Nobuhara, J. (1975). *Phytochemistry* **14,** 2304.

Okunuki, K. (1937). *Bot. Mag.* **51,** 270.

Olson, O. E. (1978). *In* "Effects of Poisonous Plants on Livestock," (R. F. Keeler, K. R. van Kampen, and L. F. James, eds.), p. 121. Academic Press, New York.

O'Neal, R. M., Chen, C.-H., Reynolds, C. S., Meghal, S. K., and Koeppe, R. E. (1968). *Biochem. J.* **106,** 699.

O'Neal, T. D. (1975). *Plant Physiol.* **55,** 975.

Owen, L. N. (1958). *Vet. Rec.* **70,** 454.

Parr, W., and Howard, P. Y. (1973). *Anal. Chem.* **45,** 711.

Partridge, S. M., and Brimley, R. C. (1952). *Biochem. J.* **51,** 628.

Pate, J. S., Walker, J., and Wallace, W. (1965). *Ann. Bot. (London)* [N.S.] **29,** 475.

Patel, N., Pierson, D. L., and Jensen, R. A. (1977). *J. Biol. Chem.* **252,** 5839.

Patrick, S. J. (1954). *J. Appl. Physiol.* **7,** 140.

Pearn, J. H. (1967a). *Br. J. Exp. Pathol.* **48,** 620.

Pearn, J. H. (1967b). *Nature (London)* **215,** 980.

Pearn, J. H., and Hegarty, M. P. (1970). *Br. J. Exp. Pathol.* **51,** 34.

Peiris, P. S., and Seneviratne, A. S. (1977). *Phytochemistry* **16,** 182.

Persaud, T. V. N. (1967). *W. Indian Med. J.* **16,** 193.

Persaud, T. V. N. (1968). *Nature (London)* **217,** 471.

Persaud, T. V. N. (1970). *Experientia* **27,** 414.

Persaud, T. V. N. (1976). *In* "Hypoglycin" (E. A. Kean, ed.), p. 55. Academic Press, New York.

Persaud, T. V. N., and Kaplan, S. (1970). *Life Sci.* **9,** 1305.

Peterkofsky, B., and Udenfriend, S. (1963). *J. Biol. Chem.* **238,** 3966.

Peterson, P. J., and Butler, G. W. (1967). *Nature (London)* **213,** 599.

Peterson, P. J., and Butler, G. W. (1971). *Aust. J. Biol. Sci.* **24,** 175.

Peterson, P. J., and Fowden, L. (1965). *Biochem. J.* **97,** 112.

Peterson, P. J., and Fowden, L. (1972). *Phytochemistry* **11,** 663.

Pfeffer, M., and Ressler, C. (1967). *Biochem. Pharmacol.* **16,** 2299.

Piez, D. A. (1968). *Annu. Rev. Biochem.* **37,** 547.

Pines, M., Rosenthal, G. A., and Applebaum, S. W. (1981). *Proc. Natl. Acad. Sci.* **78,** 5480.

Pollard, J. K., and Steward, F. C. (1959). *J. Exp. Bot.* **10,** 17.

Pollock, G. E., Oyama, V. I., and Johnson, R. D. (1965). *J. Gas Chromatogr.* **3,** 174.

Polsky, F. I., Nunn, P. B., and Bell, E. A. (1972). *Fed. Proc., Fed. Am. Soc. Exp. Biol.* **31,** 1473.

Prabhakaran, K., Harris, E. B., and Kirchheimer, W. F. (1973). *Cytobios* **7,** 245.

Prockop, D., Kaplan, A., and Udenfriend, S. (1962). *Biochem. Biophys. Res. Commun.* **9,** 162.

Qureshi, M. Y., Pilbeam, D. J., Evans, C. S., and Bell, E. A. (1977). *Phytochemistry* **16,** 477.

Rafaeli, A., and Applebaum, S. W. (1980). *Nature (London)* **283,** 872.

Rahiala, E.-L. (1973). *Acta Chem. Scand.* **27,** 3861.

Rahiala, E.-L., Kekomäki, M., Jänne, J., Raina, A., and Räihä, N. C. R. (1971). *Biochim. Biophys. Acta* **227,** 337.

Ranki, M., and Kääriäinen, L. (1969). *Ann. Med. Exp. Fenn.* **47,** 65.

Ranki, M., and Kääriäinen, L. (1970). *Ann. Med. Exp. Fenn.* **48,** 65.

Rao, S. L. N., and Sarma, P. S. (1967). *Biochem. Pharmacol.* **16,** 218.

Rao, S. L. N., Ramachandran, L. K., and Adiga, P. R. (1963). *Biochemistry* **2,** 298.

Rao, S. L. N., Adiga, P. R., and Sarma, P. S. (1964). *Biochemistry* **3,** 432.

Rao, S. L. N., Sarma, P. S., Mani, K. S., Raghunatha Rao, T., and Sriramachari, S. (1967). *Nature (London)* **214,** 610.

Ratner, S., and Rochovansky, O., (1956). *Arch. Biochem. Biophys.* **63,** 277.

Raulin, F., Shapshak, P., and Khare, B. N. (1972). *J. Chromatogr.* **73,** 35.

Rehr, S. S., Bell, E. A., Janzen, D. H., and Feeny, P. P. (1973a). *Biochem. Syst. Ecol.* **1,** 63.

Rehr, S. S., Janzen, D. H., and Feeny, P. P. (1973b). *Science* **181,** 81.

Reis, P. J. (1975). *Aust. J. Biol. Sci.* **28,** 483.

Reis, P. J., Tunks, D. A., and Chapman, R. E. (1975a). *Aust. J. Biol. Sci.* **28,** 69.

Reis, P. J., Tunks, D. A., and Hegarty, M. P. (1975b). *Aust. J. Biol. Sci.* **28,** 495.

Remmen, S. F. A., and Ellis, E. E. (1980). *Phytochemistry* **19,** 1421.

Renz, J. (1936). *Hoppe-Seyler's Z. Physiol. Chem.* **244,** 153.

Ressler, C. (1962). *J. Biol. Chem.* **237,** 733.

Ressler, C. (1975). *Recent Adv. Phytochem.* **9,** 151.

Ressler, C., and Ratzkin, H. (1961). *J. Org. Chem.* **26,** 3356.

Ressler, C., Redstone, P. A., and Erenberg, R. H. (1961). *Science* **134,** 188.

Ressler, C., Nigam, S. N., Giza, Y.-H., and Nelson, H. (1963). *J. Am. Chem. Soc.* **85,** 3311.

Ressler, C., Nelson, J., and Pfeffer, M. (1964). *Nature (London)* **203,** 1286.

Ressler, C., Nelson, J., and Pfeffer, M. (1967). *Biochem. Pharmacol.* **16,** 2309.

Ressler, C., Nigam, S. N., and Giza, Y.-H. (1969). *J. Am. Chem. Soc.* **91,** 2758.

Rhoades, D., Rendon, G. A., and Stewart, G. R. (1976). *Planta* **129,** 203.

Rhoades, D. F. (1979). *In* "Herbivores: Their Interaction with Secondary Plant Metabolites" (G. A. Rosenthal and D. H. Janzen, eds.), p. 3. Academic Press, New York.

Rockland, L. B., and Underwood, J. C. (1954). *Anal. Chem.* **26,** 1557.

Rognes, S. E. (1980). *Phytochemistry* **19,** 2287.

Rokushika, S., Murakami, F., and Hatano, H. (1977). *J. Chromatogr.* **130,** 324.

Rosenfeld, I., and Beath, O. A. (1964). "Selenium," 2nd ed. Academic Press, New York.

Rosenthal, G. A. (1970). *Plant Physiol.* **46,** 213.

Rosenthal, G. A. (1972). *Plant Physiol.* **50,** 328.

Rosenthal, G. A. (1973). *Anal. Biochem.* **51,** 354.

Rosenthal, G. A. (1974). *J. Exp. Bot.* **25,** 609.

Rosenthal, G. A. (1977a). *Q. Rev. Biol.* **52,** 155.

Rosenthal, G. A. (1977b). *Anal. Biochem.* **77,** 141.

Rosenthal, G. A. (1977c). *Biochem. Syst. Ecol.* **5,** 219.

Rosenthal, G. A. (1978a). *Life Sci.* **23,** 93.

Rosenthal, G. A. (1978b). *Experientia* **34,** 1539.

Rosenthal, G. A. (1981). *Eur. J. Biochem.* **114,** 301.

Rosenthal, G. A. (1982). *Plant Physiol.* **69,** 1066.

Rosenthal, G. A., and Bell, E. A. (1979). *In* "Herbivores: Their Interaction with Secondary Plant Metabolites" (G. A. Rosenthal and D. H. Janzen, eds.), p. 353. Academic Press, New York.

Rosenthal, G. A., and Dahlman, D. L. (1975). *Comp. Biochem. Physiol. A* **52A,** 105.

Rosenthal, G. A., and Janzen, D. H. (1981c). *Biochem. Ecol. Syst.* **9,** 219.

Rosenthal, G. A., Gulati, D. K., and Sabharwal, P. S. (1975). *Plant Physiol.* **56,** 420.

Rosenthal, G. A., Gulati, D. K., and Sabharwal, P. S. (1976a). *Plant Physiol.* **57,** 493.

Rosenthal, G. A., Dahlman, D. L., and Janzen, D. H. (1976b). *Science* **192,** 256.

Rosenthal, G. A., Janzen, D. H., and Dahlman, D. L. (1977). *Science* **196,** 658.

Rosenthal, G. A., Dahlman, D. L., and Janzen, D. H. (1978). *Science* **202,** 528.

Roth, M., and Hampaï, A. (1973). *J. Chromatogr.* **83,** 353.

Roubelakis, K. A., and Kliewer, W. R. (1978a). *Plant Physiol.* **62,** 337.

Roubelakis, K. A., and Kliewer, W. R. (1978b). *Plant Physiol.* **62,** 340.

Roubelakis, K. A., and Kliewer, W. R. (1978c). *Plant Physiol.* **62,** 344.

Roy, D. B., and Narasinga Rao, B. S. (1968). *Curr. Sci.* **37,** 395.

Rubin, J. L., and Jensen, R. A. (1979). *Plant Physiol.* **64,** 727.

Rukmini, C. (1969). *Indian J. Biochem.* **7,** 1062.

Ryan, W. L., Johnson, R. J., and Dimari, S. (1969). *Arch. Biochem. Biophys.* **131,** 521.

Sadava, C., and Chrispeels, M. J. (1971). *Biochim. Biophys. Acta* **227,** 278.

Saito, K., and Komamine, A. (1978). *Eur. J. Biochem.* **82,** 385.

Saito, K., Komamine, A., and Senoh, S. (1975). *Z. Naturforsch., C: Biosci.* **30C,** 659.

Sarma, P. S., and Padmanaban, G. (1969). *In* "Toxic Constituents of Plant Foodstuffs" (I. E. Liener, ed.), p. 267. Academic Press, New York.

Sasaoka, K., Kito, M., and Onishi, Y. (1965). *Agric. Biol. Chem.* **29,** 984.

Schachtele, C. F., and Rogers, P. (1965). *J. Mol. Biol.* **14,** 475.

Schachtele, C. F., Anderson, D. L., and Rogers, P. (1970). *J. Mol. Biol.* **49,** 255.

Schenk, W., and Schütte, H. R. (1963). *Flora (Jena, 1818–1965)* **153,** 426.

Schilling, E. D., and Strong, F. M. (1954). *J. Am. Chem. Soc.* **76,** 2848.

Schlesinger, 'H. M., Applebaum, S. W., and Birk, Y. (1976). *J. Insect Physiol.* **22,** 1421.

Schlesinger, S. (1968). *J. Biol. Chem.* **243,** 3877.

Schmitt, J. H., and Zenk, M. H. (1968). *Anal. Biochem.* **23,** 433.

Schwarz, K., and Sweeney, E. (1964). *Proc. Soc. Exp. Biol. Med.* **23,** 421.

Scott, H. H. (1917). *Ann. Trop. Med. Parasitol.* **10,** 1.

Seigler, D. S. (1977). *Biochem. Syst. Ecol.* **5,** 195.

Seigler, D. S., and Price, P. W. (1976). *Am. Nat.* **110,** 101.

Selye, H. (1957). *Rev. Can. Biol.* **16,** 1.

Seneviratne, A. S., and Fowden, L. (1968a). *Phytochemistry* **7,** 1034.

Seneviratne, A. S., and Fowden, L. (1968b). *Phytochemistry* **7,** 1047.

Senior, A. E. (1967). Ph.D. Dissertation, University of Newcastle upon Tyne.

Senoh, S., Imamoto, S., Maeno, Y., Tokuyama, T., Sakan, T., Komamine, A., and Hattori, S. (1964). *Tetrahedron Lett.* **46,** 3431.

Shargool, P. D. (1971). *Phytochemistry* **10,** 2029.

Shargool, P. D., and Cossins, E. A. (1968). *Can. J. Biochem.* **46,** 393.

Shepard, D. V., and Thurman, D. A. (1973). *Phytochemistry* **12,** 1937.

Sherratt, H. S. A., and Osmundsen, H. (1976). *Biochem. Pharmacol.* **25,** 743.

Shiman, R., Akino, M., and Kaufman, S. (1971). *J. Biol. Chem.* **246,** 1330.

Shimura, Y., and Vogel, H. J. (1966). *Biochim. Biophys. Acta* **118,** 396.

Shrift, A. (1967). *In* "Selenium in Biomedicine" (O. Muth, J. Oldfield, and P. Weswig, eds.). Avi Publ. Co., Westport, Connecticut.

Shrift, A. (1969). *Annu. Rev. Plant Physiol.* **20,** 475.

Shrift, A. (1972). *In* "Phytochemical Ecology" (J. B. Harborne, ed.), p. 145. Academic Press, New York.

Shrift, A., Bechard, D., Harcup, C., and Fowden, L. (1976). *Plant Physiol.* **58,** 248.

Siegel, R. C., and Martin, G. R. (1970). *J. Biol. Chem.* **245,** 1653.

Siegel, R. C., Pinnell, S. R., and Martin, G. R. (1970). *Biochemistry* **9,** 4486.

Smith, I. K. (1977). *Phytochemistry* **16,** 1293.

Smith, I. K., and Fowden, L. (1966). *J. Exp. Bot.* **17,** 750.

Smith, I. K., and Fowden, L. (1968). *Phytochemistry* **7,** 1065.

Smith, T. A., and Best, G. R. (1976). *Phytochemistry* **15,** 1565.

Sodek, L. (1978). *Rev. Bras. Bot.* **1,** 65.

Sodek, L., Lea, P. J., and Miflin, B. J. (1980). *Plant Physiol.* **65,** 22.

Sørensen, H. (1976). *Phytochemistry* **15,** 1527.

Spackman, D. H., Stein, W. H., and Moore, S. (1958). *Anal. Chem.* **30,** 1190.

Spener, F., and Dieckhoff, M. (1973). *J. Chromatogr. Sci.* **11,** 661.

Stadtman, T. C. (1974). *Science* **183,** 915.

Stark, R. (1967). *In* "Methods in Enzymology" (C. H. Werner Hirs, ed.), Vol. 11, p. 125. Academic Press, New York.

Stein, S., Bohlen, P., Stone, J., Dairman, W., and Udenfriend, S. (1973). *Arch. Biochem. Biophys.* **155,** 202.

Steward, F. C., and Durzan, D. J. (1965). *In* "Plant Physiology: A Treatise" (F. C. Steward, ed.), Vol. 4A, p. 451. Academic Press, New York.

Steward, F. C., Bidwell, R. G. S., and Yemm, E. W. (1956). *Nature (London)* **178,** 734.

Stewart, G. R., and Rhoades, D. (1977). *New Phytol.* **79,** 257.

Strassman, M., and Weinhouse, S. (1953). *J. Am. Chem. Soc.* **75,** 1680.

Streeter, J. G. (1977). *Plant Physiol.* **60,** 235.

Stuart, K. L. (1976). *In* "Hypoglycin" (E. A. Kean, ed.), p. 39. Academic Press, New York.

Stuben, I., Israelstam, G. F., and Oaks, A. (1979). *Planta* **146,** 237.

Su, E. F. W., and Levenberg, B. (1967). *Acta Chem. Scand.* **21,** 493.

Suda, S. (1960). *Bot. Mag.* **73,** 142.

Suharjono and Sadatun (1968). *Paediatr. Indones.* **8,** 20.

Sung, M.-L., and Fowden, L. (1969). *Phytochemistry* **8,** 2029.

Sung, M.-L., and Fowden, L. (1971). *Phytochemistry* **10,** 1523.

Sung, M.-L., Fowden, L., Millington, D. S., and Sheppard, R. C. (1969). *Phytochemistry* **8,** 1227.

Suttle, J. C., and Kende, H. (1980). *Phytochemistry* **19,** 1075.

Suzuki, T., Komatsu, K., and Tuzimura, K. (1973). *J. Chromatogr.* **80,** 199.

Synge, R. L. M. (1968). *Annu. Rev. Plant Physiol.* **19,** 113.

Takahashi, M., and Ripperton, J. C. (1949). *Tech. Bull.—Hawaii Agric. Exp. Stn.* **100.**

Takemoto, T., Nomoto, K., Fushiya, S., Ouchi, R., Kusano, G., Hikino, H., Takagi, S., Matsuura, Y., and Kakudo, M. (1978). *Proc. Jpn. Acad., Ser. B.* **54,** 469.

Takeuchi, T., and Prockop, D. J. (1969). *Biochim. Biophys. Acta* **175,** 142.

Tanaka, H., and Stadtman, T. C. (1979). *J. Biol. Chem.* **254,** 477.

Tang, S.-Y., and Ling, K.-H. (1975). *Toxicon* **13,** 339.

Tate, S. S., and Meister, A. (1973). *In* "The Enzymes of Glutamine Metabolism" (S. Prusiner and E. R. Stadtman, eds.), p. 77. Academic Press, New York.

Terano, S., and Suzuki, Y. (1978). *Phytochemistry* **17,** 550.

Thanjan, D. K. (1967). M. S. Thesis, University of Hawaii, Honolulu.

Thompson, J. F. (1980). *In* "The Biochemistry of Plants" (B. J. Miflin, ed.), Vol. 5, p. 375. Academic Press, New York.

Thompson, J. F., and Moore, D. P. (1967). *Biochem. Biophys. Res. Commun.* **28,** 474.

Thompson, J. F., and Moore, D. P. (1968). *Biochem. Biophys. Res. Commun.* **31,** 281.

Thompson, J. F., and Morris, C. J. (1968). *Arch. Biochem. Biophys.* **125,** 362.

Thompson, J. F., Turner, D. H., and Gering, R. K. (1964). *Phytochemistry* **3,** 33.

Tiwari, H. P., Penrose, W. R., and Spenser, I. D. (1967). *Phytochemistry* **6,** 1245.

Tixier, M., and Desmaison, A. M. (1980). *Phytochemistry* **19,** 1643.

Tong, J. H., D'Iorio, A., and Benoiton, N. L. (1971). *Biochem. Biophys. Res. Commun.* **43,** 819.

Toome, V., and Reymond, G. (1975). *Biochem. Biophys. Res. Commun.* **66,** 75.

Töpfer, V. R., Miersch, J., and Reinbothe, H. (1970). *Biochem. Physiol. Pflanzen.* **161,** 231.

Torquati, T. (1913). *Arch. Farmacol. Sper. Sci. Affini* **15,** 308.

Touchstone, J. C., Sherma, J., Dobbins, M. F., and Hansen, G. R. (1976). *J. Chromatogr.* **124,** 111.

Tschiersch, B. (1964). *Phytochemistry* **3,** 365.

Udenfriend, S., Titus, E., Weissbach, H., and Peterson, R. E. (1956). *J. Biol. Chem.* **219,** 335.

Udenfriend, S., Weissbach, H., and Bogdanski, D. F. (1957). *J. Biol. Chem.* **224,** 803.

Vanderzant, E. S., and Chremos, J. H. (1971). *Ann. Entomol. Soc. Am.* **64,** 480.

van Etten, C. H., and Miller, R. W. (1963). *Econ. Bot.* **17,** 107.

van Veen, A. G. (1973). *In* "Toxicants Occurring Naturally in Foods," 2nd ed., p. 464. Natl. Acad. Sci., Washington, D.C.

van Veen, A. G., and Latuasan, H. E. (1949). *Indones. J. Nat. Sci.* **105,** 288.

Vega, A., and Bell, E. A. (1967). *Phytochemistry* **6,** 759.

Vega, A., Bell, E. A., and Nunn, P. B. (1968). *Phytochemistry* **7,** 1885.

Virtanen, A. I., and Linko, P. (1955). *Acta Chem. Scand.* **9,** 531.

Virtanen, A. I., and Matikkala, E. J. (1960). *Hoppe-Seyler's Z. Physiol. Chem.* **322,** 8.

Virupaksha, T., and Shrift, A. (1963). *Biochim. Biophys. Acta* **74,** 791.

Virupaksha, T. K., and Shrift, A. (1965). *Biochim. Biophys. Acta* **107,** 69.

von Holt, C., and von Holt, L. (1958). *Naturwissenschaften* **45,** 546.

von Holt, C., von Holt, L., and Böhm, H. (1966). *Biochim. Biophys. Acta* **125,** 11.

Wada, M. (1930). *Biochem. Z.* **224,** 420.

Waley, S. G. (1966). *Adv. Protein Chem.* **21,** 1.

Walker, J. B. (1955a). *J. Biol. Chem.* **212,** 207.

Walker, J. B. (1955b). *J. Biol. Chem.* **212,** 617.

Ward, K. A., and Harris, R. L. N. (1976). *Aust. J. Biol. Sci.* **29,** 189.

Watkins, J. C., Curtis, D. R., and Biscoe, T. J. (1966). *Nature (London)* **211,** 637.

Watson, R., and Fowden, L. (1973). *Phytochemistry* **12,** 617.

Wayman, O., and Iwanaga, I. I. (1957). *Hawaii, Agric. Exp. Stn., Misc. Publ.* **81.**

Weaks, T. E., Jr., and Hunt, G. E. (1973). *Physiol. Plant.* **29,** 421.

Weigele, M., DeBernardo, S., and Leimgruber, W. (1973). *Biochem. Biophys. Res. Commun.* **50,** 352.

Whittaker, R. H., and Feeny, P. P. (1971). *Science* **171**, 757.

Wiebers, J. L., and Garner, H. R. (1963). *Abstr., 145th Meet., Am. Chem. Soc.*, p. 22C.

Wilding, M. D., and Stahmann, M. A. (1962). *Phytochemistry* **1**, 263.

Willet, E. L., Henke, L. A., Quisenberry, J. H., and Maruyama, C. (1947). *Bienn. Rep.—Hawaii Agric. Exp., Bienn. Rep., 1944–1946*.

Williams, B. L., and Wilson, K. (1975). *In* "Principles and Techniques of Practical Biochemistry" (B. L. Williams and K. Wilson, eds.), p. 52. Elsevier, New York.

Index